D0897717

After Geoengineering

After Geoengineering

*Climate Tragedy,
Repair, and Restoration*

Holly Jean Buck

VERSO
London • New York

First published by Verso 2019
© Holly Jean Buck 2019

1 3 5 7 9 10 8 6 4 2

Verso
UK: 6 Meard Street, London W1F 0EG
US: 20 Jay Street, Suite 1010, Brooklyn, NY 11201
versobooks.com

Verso is the imprint of New Left Books

ISBN-13: 978-1-78873-036-5
ISBN-13: 978-1-78663-799-4 (UK EBK)
ISBN-13: 978-1-78873-038-9 (US EBK)

British Library Cataloguing in Publication Data
A catalogue record for this book is available from the British Library

Library of Congress Cataloging-in-Publication Data
A catalog record for this book is available from the Library of Congress

Typeset in Fournier by MJ & N Gavan, Truro, Cornwall
Printed and bound by CPI Group (UK) Ltd, Croydon, CR0 4YY

Contents

Introduction

Desperation Point

December in California at one degree of warming: ash motes float lazily through the afternoon light as distant wildfires rage. This smoky "winter" follows a brutal autumn at one degree of warming: a wayward hurricane roared toward Ireland, while Puerto Rico's grid, lashed by winds, remains dark. This winter, the stratospheric winds break down. The polar jet splits and warps, shoving cold air into the middle of the United States. Then, summer again: drought grips Europe, forests in Sweden are burning, the Rhine is drying up. And so on.

One degree of warming has already revealed itself to be about more than just elevated temperatures. Wild variability is the new normal. Atmospheric patterns get stuck in place, creating multi-week spells of weather that are out of place. Megafires and extreme events are also the new normal—or the new abnormal, as Jerry Brown, California's former governor, put it. One degree is more than one unit of measurement. One degree is about the uncanny, and the unfamiliar.

If this is one degree, what will three degrees be like? Four?

At some point—maybe it will be two, or three, or four degrees of warming—people will lose hope in the capacity of current emissions-reduction measures to avert climate upheaval. On one hand, there is a personal threshold at which one loses hope: many of the climate scientists I know are there already. But there's also a societal threshold: a turning point, after which the collective

discourse of ambition will slip into something else. A shift of narrative. Voices that say, "Let's be realistic; we're not going to make it."
Whatever *making it* means: perhaps limiting warming to 2°C, or 1.5,
as the Paris Agreement urged the world to strive for. There will be
a moment where "we," in some kind of implied community, decide
that something else must be tried. Where "we" say: Okay, it's too
late. We didn't try our best, and now we are in that bad future. Then,
there will be grappling for *something* that can be done.

This is the point where it becomes "necessary" to consider the
future we didn't want: solar geoengineering. People will talk about
changing how we live, from diet to consumption to transportation;
but by then, the geophysics of the system will no longer be on our
side. A specter rears its head: the idea of injecting aerosols into the
stratosphere to block incoming sunlight. The vision is one of shield-
ing ourselves in a haze of intentional pollution, a security blanket
that now seems safer than the alternative. This discussion, while not
an absolute given, seems plausible, if not probable, from the vantage
point of one degree of warming—especially given that emissions
are still rising.

You may have heard something about solar geoengineering. It's
been skulking in the shadows of climate policy for a decade, and
haunting science for longer than that, even though it's still just a
rough idea. But it is unlikely you imagined solar geoengineering
would be a serious topic of discussion, because it sounds too crazy—
change the reflectivity of the earth to send more sunlight back out
into space? Indeed, it *is* a drastic idea.

We are fortunate to have rays of sunlight streaming through space
and hitting the atmospheric borders of our planet at a "solar constant"
of about 1,360 watts per square meter (W/m^2) where the planet is
directly facing the sun. This solar constant is our greatest resource;
a foundation of life on earth. In fact, it's not actually so constant
—it was named before people were able to measure it from space.
The solar constant varies during the year, day to day, even minute to
minute. Nevertheless, this incoming solar energy is one of the few
things in life we can count on.

Much of this sunlight does not reach the surface; about 30 percent of it gets reflected back into space. So on a clear day when the sun is at its zenith, the solar radiation might reach 1,000 W/m^2. But this varies depending on where you are on the globe, on the time of day, on the reflectivity of the surface (ice, desert, forest, ocean, etc.), on the clouds, on the composition of the atmosphere, and so on. Because it's night half the time, and because the sun is hitting most of the earth at an angle, the average solar radiation around the globe works out to about 180 W/m^2 over land.[1] Still, this 180 W/m^2 is a bounty.

The point of reciting all these numbers is this: solar geoengineering amounts to an effort to change this math. That's how a researcher might look at it, anyway.

From one perspective, it sounds like complete lunacy to intentionally mess with something as fundamental as incoming solar radiation. The sun, after all, has been worshipped by cultures around the world: countless prayers uttered to Ra, Helios, Sol, Bel, Surya, Amaterasu, and countless other solar deities throughout the ages. Today, many still celebrate holidays descended from solar worship. And that worship makes sense—without the sun, there would be nothing. Even in late capitalism, we valorize the sun: people search for living spaces with great natural light; they get suntans; they create tourist destinations with marketing based on the sun and bring entire populations to them via aircraft. Changing the way sunlight reaches us and *all other life on earth* is almost unimaginably drastic.

But there are ways of talking about solar geoengineering that normalize it, that make you forget the thing being discussed is sunlight itself. The most discussed method of solar geoengineering is "stratospheric aerosol injection"—that is, putting particles into the stratosphere, a layer of the atmosphere higher than planes normally fly. These particles would block some fraction of incoming sunlight, perhaps about 1 to 2 percent of it. Stratospheric aerosols would change not only the amount of light coming down, but also the type: the light would be more diffuse, scattering differently. These changes would alter the color of our skies, whitening them to a degree that may or may not be easily perceptible, depending on whether you live in an urban area. The distortion would also affect how plants and phytoplankton operate. Certainly, this type of inter-vention seems extreme.

And despite the extremity of the idea, it's not straightforwardly irrational. First of all, solar radiation is already naturally variable; a single passing cloud can change the flux by 25 W/m².[2] What's more, solar radiation is *unnaturally* variable. Global warming is caused by greenhouse gas emissions—the greenhouse gas molecules trap heat, creating an imbalance between the energy coming in and the energy going back out. Since 1750, these emissions have increased the flux another 2.29 W/m².[3] This disparity between incoming and outgoing energy is what scientists call "radiative forcing"—a measure of imbalance, of forced change, caused by human activity. That imbal-ance would actually be greater—just over 3 W/m²—if not for the slight countervailing effect of aerosol emissions that remain close to the ground. Think about a smoggy day. The quality of the light is dimmer. Indeed, air pollution from cars, trucks, and factories on the ground already masks about a degree of warming. Total removal of aerosols—as we're trying to accomplish, in order to improve air quality and human health—could induce heating of 0.5 to 1.1°C globally.[4]

So, from another perspective, because human activity is already messing with the balance of radiation through both greenhouse gas emissions (warming) and emitting particulate matter from industry and vehicles (cooling), it doesn't sound as absurd to entertain the idea that another tweak might not be that significant—especially if the counterfactual scenario is extreme climate suffering. If you stretch your imagination, you can picture a future scenario where it could be more outrageous *not* to talk about this idea.

The question is, are we at the point—let's call it "the shift"—where it is worth talking about more radical or extreme measures—such as removing carbon from the atmosphere, leaving oil in the ground, social and cultural change, radical adaptation, or even solar geoengineering?

Deciding where the shift—the moment of reckoning, the desper-ation point—lies is a difficult task, because for every optimist who thinks renewables will save the day, there is a pessimist noting that the storage capacity and electrical grid needed for a true renewable revolution does not even exist as a plan. For many people, it's hard to tell how desperate to feel: we know we should be worried, but we also imagine the world might slide to safety, show up five minutes to midnight and catch the train to an okay place, with some last-minute luck. It can seem like the dissonance around what's possible actually *increases* the closer we get to the crunch point; the event horizon. Some of this uncertainty is indeed grounded in the science. "Climate sensitivity"—the measurement describing how earth would respond to a doubling of greenhouse gas concentrations from preindustrial times—is still unknown. That means we don't know precisely what impacts a given amount of greenhouse gas emissions will have.

However, basic physics dictates that this season of uncertainty is limited. The picture will become clearer as emissions continue, and as scientists tally up how much carbon is in the atmosphere. Nevertheless, examining the situation today provides useful insights that should be well known, but somehow are rarely discussed in venues other than technical scientific meetings.

At present, human activities emit about 40 gigatons (Gt) of carbon dioxide a year, or 50 Gt of "carbon dioxide equivalent," a measure that includes other greenhouse gases like methane. (A gigaton is a billion tons.) Since the Industrial Revolution, humans have emitted about 2,200 Gt of CO_2.[5] Scientists have estimated that releasing another 1,000 Gt CO_2 equivalent during this century would raise temperatures by two degrees Celsius—exceeding the target of the Paris Agreement—meaning that 1,000 Gt CO_2 is, if you like, our maximum remaining budget (these are rough figures; it could be much less).[6] Knowing that today roughly 50 Gt of carbon dioxide equivalent is emitted, it is evident that emitters are on track to squander the entire carbon budget within the next 20 years. Moreover, the *rate* of warming is still increasing. This means that if the rate of warming slows down yet emissions remain at today's rate, in twenty years, two degrees of warming are essentially guaranteed.

What would it take to avoid this? To keep warming below two degrees, emissions will need to drop dramatically—and even go negative by the end of this century, according to scenarios assessed by the Intergovernmental Panel on Climate Change. Figure 1 shows a typical "okay future" scenario; one that would provide for a decent chance of staying within two degrees.

Three key features are evident in Figure 1.

First, the "good future" scenario has emissions peaking around 2020, and then dropping dramatically. Dramatic emissions reductions are key to any scenario that limits warming.

Second, emissions go net negative around 2070. "Net negative" means that the world is sucking up more carbon than it is emitting. How is that done? While emissions can be zeroed via the mitigation measures we're familiar with—using renewable energy instead of fossil fuels, stopping deforestation, halting the destruction of wetlands, and so on—to push emissions beyond zero and into negative territory requires a greater degree of intervention. There are two main categories of approach: biological methods, including using forests, agricultural systems, and marine environments to store carbon; and geologic methods, which typically employ industrial means to capture and store CO_2 underground or in rock. Some approaches combine these, though: for instance, coupling bioenergy

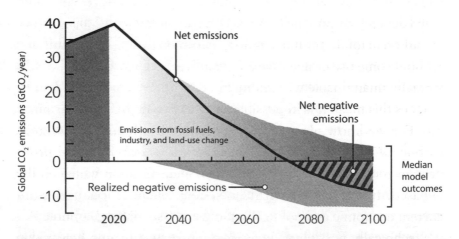

Figure 1. *Median values from 18 scenarios evaluated by six models using shared socioeconomic pathways assessed in the next assessment report of the Intergovernmental Panel on Climate Change. Data: Glen Peters / CICERO*

with carbon capture and storage. (We'll walk through these different practices in detail later.)

But here, note a third point: carbon actually *starts* to be removed in the 2020s and 2030s, when emissions are still relatively high. Industrial carbon capture and storage (CCS)—the practice of capturing streams of carbon at industrial sites and injecting it into underground wells—is a crucial technique for accomplishing these levels of carbon removal. As of 2019, the world has only around twenty CCS plants in operation, a number that is almost quaint in scale. To begin removing carbon at the level displayed in Figure 1 implies scaling up the current amount of carbon stored by something like a thousandfold. By 2100, in this scenario, the world would be sequestering ten or fifteen gigatons of carbon dioxide equivalent. And the scale-up begins right away.

The gentle slope of declining greenhouse gases looks so neat and calm. It is a fantasy described in clean lines; in the language of numbers, the same language engineers and builders and technocrats speak. This language lends weight to the image, making it seem less fantastic. However, this scenario relies on carbon removal technology at a scale far beyond the demonstration projects being planned today. As the Intergovernmental Panel on Climate Change (IPCC) warned in its special report on 1.5°C, reliance on such technology is a major risk. But the same report indicated that all the pathways analyzed depended upon the removal of between 100 and 1,000 gigatons of carbon in total.[7] In short, limiting warming to 2°C is very difficult without some use of negative emissions technologies—and 1.5°C is virtually unattainable without them.

Does this mean it is impossible to avert two degrees of warming? No. For we know plenty of practices that can be used to remove carbon. We can store it in soils, in building materials and products, in rock. After all, it's a prevalent element upon which all life is based. It would be difficult to scale these practices under our current economic and political logic, as we'll explore this book. But it's technically possible to imagine a future like the one depicted in Figure 1—a future where the excesses of the past (our present) are tucked away, cleaned up, like a stain removed. Some call this vision "climate restoration": the idea that we could use carbon removal technologies and practices to draw down carbon dioxide concentrations from the current 410 parts per million (ppm), down to the 300s, or even back to preindustrial levels. This vision undoubtedly strikes many as grandiose, and perhaps as unnecessary: after all, in many economies in the global North, emissions look to be plateauing or even declining.

If emissions *do* peak in the next decade, does that mean we're safe?

2030s

You're sitting at the kitchen table. It's hot. The radio says: "... and in Bonn today, negotiators hammered out a new agreement on differentiated responsibilities for carbon removal. We're also hearing rumors of a blockchain-based system of accounting to track international carbon flows.

"Meanwhile, at the Detroit Auto Show, yet another manufacturer has announced that their fleet will be all-electric by 2032. Following the rollout of Tesla's heavy trucks, Škoda has unveiled the design ..."

You turn to your tablet, scrolling through your feeds as you sip your coffee. A picture of bright-red apples harvested from the student-run orchard: you press Like. Someone has posted an article with the headline "Emissions Peak Is Only a Plateau," subtitled: "The Low-Hanging Mitigation Fruit Has Been Picked—Now the Hard Work Begins." You want to read it and comment, so that the poster will know you're smart enough to follow what they post— but it sounds like a downer.

If you think the actions on climate will basically work out, turn to the next page.

If you think it's all just talk, performance, showmanship, turn to page 10.

2040s

Ten years, five months, and eighteen days later, you walk over to the kitchen table with fresh slices of slightly burnt toast. Your spouse doesn't look up. "What are you reading?" you ask.

"It's an article that looks at our success with climate change, and talks about how to apply it to public health and other global challenges." They chomp on the toast, leaving crumbs on the table. "We actively phased out fossil fuel infrastructure, and improved energy efficiency. Afforestation, carbon farming, and wetland restoration policies were also huge. Historically unprecedented rates of land-use change to sequester carbon, three times faster than the conversion of forests to soybeans in previous decades."

"Mm-hmm," you say. "Thank god the ag companies invested in climate change–resistant coffee strains." You have that feeling you get when it is late in the morning and you haven't accomplished very much yet.

"There was also the successful push toward carbon capture and storage. Back in 2017, they had only twenty-some CCS plants. But in 2020, with the global initiative on negative emissions, countries were putting in CCS facilities all over the place—and now there are over 2,000. In just a few decades."

"That's great," you say.

"The article is pretty techno-focused. It was really the social movements that created the push to hold the fossil fuel companies accountable, and that changed the narrative. I mean, when we were teenagers, everything was so apocalyptic. It was like some new climate disaster was in the air every day. Our kid isn't even going to know what that's like."

Turn to page 24.

2040s

You roll over in bed, throw off the faded sheet, and think about sitting up. Your kid is still asleep. *Thank god*. You ask: "Alexa, what do I need to know?"

"It's six twenty-three. Would you like your headlines?"

"Sure," you say.

"A new paper in *Science* indicates that the world has warmed two degrees since the beginning of the Industrial Revolution. This threshold was crossed on the heels of the ongoing water emergency in Yemen. Meanwhile, the US House of Representatives is voting today on whether to join other countries in a geoengineering accord. Would you like to hear more of the story?"

"Sure," you say, pulling on your socks.

"While emissions have slightly decreased in several regions, including the United States and China, they are still rising rapidly in other parts of the world. There's disagreement around whether or not emissions have plateaued, and whether we are headed for 4°C of warming by the end of the century."

"The geoengineering accord would establish an international cost-sharing agreement and insurance scheme that pools international resources for the deployment of stratospheric aerosols. Florida senator Jackie Gonzalez has been the first to speak out in favor of the accord. 'If the melting of Greenland continues, Miami will be completely inundated in a few decades. It's already impossible to insure the property along our coasts. The world has offered us a reasonable path to salvation, one that will be mutually beneficial for coastal cities all around the world. This accord could preserve geopolitical stability abroad as well as protect our homeland.' However, several key members of the US Congress have remained silent. We can expect the results to be challenged in the courts either way."

You head for the kitchen table and pick up your tablet, even though you've made a resolution not to look at it first thing in the morning anymore—anyway, it's not really first thing if you're out of bed. You have three notifications from friends, all auto-sent from an environmental NGO, asking you to call your representatives to protest the geoengineering accord. *We need polluters to pay for climate cleanup—not new risks for ourselves and our kids.* The photo of the smiling family looks like it could have been shot in your neighborhood. *Sigh*—the email is one of those that is custom-tailored based on your social media profile, which means there'll be nostalgic nods to the protest activities of your teenage years, too, and pictures of your younger self. *Remember when you marched on Washington for climate justice in '27? Now that you're a parent, you've got even more to fight for! Put fossil fuels out of business. Stop solar geoengineering NOW!*

If you choose to call your representatives in protest, flip a coin. Heads: turn to page 12; tails: turn to page 13.

If you think the geoengineering accord sounds like a pretty good option, turn to page 15.

2050s

You take your sandwich out of its wrapper: a few slices of cultivated meat and a flat tomato from your building's greenhouse.

You stay inside for lunch. A few die-hard coworkers are eating on the benches outside, but they're going to return to the office drenched in sweat. The downside of your choice is that there is a TV blaring. Today's lunchtime news talk show is about negative emissions.

"Whose negative emissions are these, really?" asks the moderator. "The ecologist from Nature Conservation says that they're her negative emissions, because she's worked with landowners in Iowa for over a decade to remove this carbon. But it seems like you, Ron, are saying that they're *your* negative emissions—that her work happens so that you at LowC Fuels can produce more fuel."

"In a sense, yes. But we're doing this in service of society." Ron leans back in his chair. "Look, Jim, these negative emissions can't *belong* to anyone, because by definition they don't exist. They're negative. And it's our job to keep these airplanes flying. That means we have to emit a little, and remove a little."

The ecologist cuts in. "Well, I didn't spend my career in the fields, teaming up with hardworking ranchers to reform farm practices, so that you could continue to pile up profits. Not after your industry spent *decades* obscuring the *very existence* of climate change."

So far, the on-screen needle recording audience sentiment has favored the energy spokesperson, who has taken up the voice of sensibility; but now it begins to tilt in favor of the ecologist. You pull out your phone to send in your sentiment. You get distracted by images of drowned refugee corpses in the Mediterranean, and then by the liquid eyes of a now-extinct species of forest mouse. It's a good thing climate sensitivity turned out to be low and those Antarctic glaciers have been so hardy, you think, polishing off your minimal sandwich.

Turn to page 24.

2050s

You eat your sandwich at your desk. Your terminal is monitoring you to make sure you eat your sandwich in a reasonable amount of time.

It is the anniversary of your child's death. On your desk is a picture of her, standing on the shore, clutching a red balloon. It was taken before the hurricane hit.

You see her when she had dengue fever. Waiting in the hospital lines, sleeping in your arms, breathing hard. *You should have done something more.* Your chatbot therapist tells you these thoughts are unhelpful, but you can't exorcise them.

You crumple up the wrapper from your sandwich and sweep the crumbs into the trash. The face of your chatbot therapist appears on the screen. "Good afternoon. It looks like your blood pressure has risen somewhat. Are you feeling okay?"

"Must be a nanosensor glitch."

She blinks at you, slowly. "Would you like to talk about it?"

"I'm just thinking that we should have done something more."

"I'm not sure what you mean. Can you tell me more?"

"I mean, we should have stopped the decline before it got this bad and everyone started dying all around us."

"It's not your fault, you know."

"That's what you always say, CB. You should try to mix it up."

"Okay. Thanks for your suggestion. I've noted your response." The therapist blinks again, attentively. You've observed that she is programmed with at least six types of blinks. "You know, a lot of other people are feeling bad about climate change too. Thirty-six percent of them have talked with me this week about climate change."

"I know."

"Maybe it would help to go to the exercise room and video chat with one of them."

You sigh. Once the therapist suggests this, you normally have to do it, because they notice if you ignore the suggestion too often. "I just feel like if there was anything I could do to stop it, I would. To put things back."

"There wasn't anything that anyone could have done. At the climate emergency summit in 2048, all the options were reviewed. Climate sensitivity was high. Last-ditch options like solar geoengineering were found to be too risky and contentious. The best we can do is keep on going. Your loved ones would want that."

"Nobody wanted the liability," you mutter. "Bunch of cowards. Fuck it. I'm going to the exercise room."

Your therapist brightens her concerned eyes and blinks energetically. "Good idea!"

Turn to page 24.

2050s

You're in the doctor's waiting room. On screen, the first planes are going up. No one is paying attention; everyone's kids are miserable. Your child's breathing is slow and heavy. She won't even take a bite of the sandwich you made her.

The screen cuts to a reporter on a college campus: buildings of steel and glass, upon a patch of brown, withered grass. "The provisions in the accord say that we have to stick to the schedule for drawing down carbon, or we'll be stuck with this forever. Let's hear the word on the street."

She turns to two young men. "This is Dave and Jason. Guys, the geoengineering program starting today says that your generation, and your kids' generation, are going to be responsible not just for cutting carbon emissions, but for engaging in carbon management."

"Carbon management? Is that a major? I'm here for virtual game design," one boy says.

His friend shakes his head. "Come on. Carbon management is, like, waste management. Sure. Why not?"

The reporter nods. "We're talking about using even more technologies that remove carbon, because if we don't, we'll have to keep up the solar geoengineering forever. What do you think?"

"If you can make it a game, I'm in," the first boy says, smiling to the camera and the interviewer.

His friend pauses. "I think it sounds like a lot of responsibility, and, you know, they're leaving us with a lot of debt and not much social security. Like my mom was supposed to be getting social security, but it ran out, so she has to live with my sister. When I finish school, it'll be my turn to take her in, and I gotta find a place with two whole rooms that I can afford. So I don't know how we're supposed to have the resources to manage all this carbon and stuff. We can't even get good jobs. Our parents don't even have houses. So I don't know about that."

You turn away, stroke your fevered kid's damp hair.

If you think that the world is finally going to draw down the carbon, continue on the following page.

If you're not sure it will, turn to page 19 or 22.

2070s

"No more stories," you tell your grandson and granddaughter. "It's really time for bed."

"I want to hear 'Amelia and the Carbon Monster' next," your grandson insists, struggling with his purple pajamas.

"Okay, just one more story." You settle onto the pillows and the children clamber around you, all limbs and angles. "Amelia and the Carbon Monster," you announce, taking a deep breath.

"Amelia was walking to school one day when she found a box on the sidewalk. She opened up the box, and there was a pair of magic glasses!" You have read this story so many times you could sleep-walk through it: Amelia, with her magic glasses, can see the invisible carbon monster. Amelia pulls together her team of Special Investi-gators. They devise all these ways to dissolve the carbon monster: they plant the 10 billion trees, build the beautiful new cities from special wood, and design the machines that take carbon out of the air and put it deep underground. Meanwhile, the Scientific Commit-tee builds the Stratospheric Shield, so that the team has time to do their work.

"By the time they were finished, they were old, and the play-ground where they used to meet had become a meadow. Amelia took out her magic glasses and looked around. The trees and dirt and plants and buildings were all filled up with carbon, glowing with a soft light. The end." You close the book. Your feet have fallen asleep, and your grandson has a glassy-eyed look.

Your granddaughter still looks somewhat alert. "That's not a true story, is it?"

"It's not exactly true. But it's true in spirit," you say.

"How can carbon be a monster if it's everywhere?"

"Well, carbon's not a monster. The people who wanted to burn too much of it and get rich were kind of like monsters. The point of the book is the teamwork that had to go into cleaning up the carbon pollution. Even though it's not all cleaned up yet, we're halfway there. Pretty soon we might be able to take down the Stratospheric Shield."

"I don't like cleaning up."

"Yeah. But this is about changing your attitude. Cleaning up can be really fun, if you do it with your friends," you explain. "You know when you do cleanup time at school, and it's like a race? It can be like that. You were really my little cleanup helper when you were two."

Your granddaughter has lost interest. "I want to play with my mini forest now."

"No," you say, "it's time for bed." They will sleep well.

Turn to page 24.

2070s

You're getting ready for bed when there's a knock on your door. You tap on the wall to call up video from the hall: a professional-looking duo in bright sky-blue suits. They can't be Mormons, you think, with suits that color.

"Good evening!" They are beaming. "We're sorry to stop by so late," the woman apologizes, "but this was predicted to be a good time."

"What can I do for you?"

"We're with a group called Blue Sky Again. Do you remember the color the sky used to be when you were a kid?"

"Sure I do. But how would you? You're too young."

"We didn't grow up with true-blue skies," the man says. "But I got the blue sunglasses when they came out. We try to connect with people who do remember, and we'd like to give you some information about how you can help."

You sigh.

"You probably remember when the planes started going up."

"Yeah."

"Well, maybe you also remember the great enthusiasm for carbon removal around the time of the accord."

"Sure, the planes were supposed to be temporary until we got the carbon out. That was the line. The world cut emissions by half, and decided that was good enough, I guess." Your legs ache from the exercise machine and you begin to wonder if you should have opened the door.

"Well, we think that's a crime. I see on the wall you have some pictures of little ones—grandkids? We think they should be able to see blue skies, too. We might never be able to, but they should. So we need to finish the job."

You laugh. "That's a very laudable cause for you to spend your evenings on. But why would people start on carbon removal again? And how?"

"I'm glad you asked the 'how' question. Our five-point plan, FINISH, answers that. We can project it on the wall."

Bullet points appear beside your door. "The first step is Fund. They skipped this fifty years ago, so they never even got to the next steps. But now we have the Machine Labor Act. You're familiar with that?"

"Well, I think so. Companies that automate have to put revenue into the Universal Basic Income program, proportional to how many jobs they automated. In order to spread the benefits of automation around a bit more."

"Yes. We want to use that same model for funding further carbon removal. We can't tax only the energy companies who combusted fossil fuels; after all, they're the ones now providing our low-carbon fuels and carbon storage, and we need them to continue operating. But we do have a lot of wealth that was created in the tech sector during this century. What we're proposing to do is to add a small additional tax on that wealth. It will hurt the bottom lines of the old mega-platforms, the old Apple-Amazon-Alphabets. But not too much—and many of these companies will get it back, if they choose to participate in carbon removal activities. They get a nanobonus for each ton removed. That's our second step: Incentivize."

"I'm not sure how any of the mega-platforms are really going to remove carbon themselves, though."

"We know you grew up thinking of these as technology companies. But they've moved into transport, building, manufacturing, and all other sectors—you can see on this chart that these top four companies and their subsidiaries are responsible for a huge share of energy consumption. So there's a lot they can do, and they're the ones that have the revenue to fund carbon-negative infrastructure."

You feel like you've been standing in your doorway for a very long time.

"So, it's Fund, Incentivize—those two happen at the national scale. Then, to take this global, we need Negotiate, Inspire, and Show Heart. Negotiation will work if we inspire other countries with our model and build goodwill. It's really our responsibility to make the first move. So, Negotiate, Inspire, and Show Heart. That final step is really important. The reason this didn't work before is because it was technocratic, and unemotional. But the human heart is really at the core of Blue Sky Again."

"I'm not sure what I can do about all of this," you tell her. "Honestly, I was just getting ready for bed ..."

"It's that first step, Fund, that needs your help. These companies aren't going to fund the removal projects unless citizens speak out and show your demand for this. So we're asking you to join the movement. There are three concrete things you can do: you can talk with your elected representatives, share the key messages of FINISH, and vote."

You sigh. "I did all those things fifty years ago, and it didn't do a damn thing. And it would have been so much easier to fix then, before we'd put another several hundred gigatons in. We're addicted to these particles now." They both look a bit crestfallen. "I'm glad there's young people like you who care. But honestly, the world just doesn't work that way."

Turn to page 24.

2070s

You're crawling into bed and hear a distant blast; the windows vibrate. You get up and secure the sheeting taped over all the windows and doors. Then, you check your phone. There's the alert: *Residents within one hundred miles of Active Participation airstrips, shelter in place.*

Are you ok? It's a message from your child, 3,000 miles away.

Ok. Looks like I'll be indoors for a while, you write back. You hope the winds are going out to sea.

I wish you had evacuated and come over. You had considered it, but it would be a long journey for someone in your condition. It would also have exhausted the last of your savings, making you utterly dependent upon your child.

Then another message: *I can't believe they would do this.*

As they were growing up, you tried to cultivate a sense of hope in your child, hope for positive action and rationality—but they ended up being pretty naive, you reflect. The solar geoengineering program was believed to be a common good, out of reach of the low-level tangle of proxy wars. It was well known that stopping the program would cause a rapid temperature rise of three or four degrees over the next few years, decimating food production. Mutually assured destruction, your parents had called it, and they seemed pretty comfortable with the idea that this was a workable arrangement. However, key demographics in both Russia and the US had been fed a steady, micro-targeted stream of propaganda: ending the program would cripple their enemies while making Russia and America glorious breadbaskets again. In the past month, verified videos showed rallies in the streets of Murmansk calling for leaders to end the program for good. Now it appears this idea has buy-in from the Russian regime, whose erratic incompetence makes you long for the calculating petro-oligarchs of your youth.

They don't live in the same reality as we do, honey, you write back. You wonder how long it would take to come to a sensible agreement and to restore the flights. Conventional wisdom used to be that you had a couple of years before the last of the particles fell out and the warming began—plenty of time to manufacture more planes and set up more airstrips. A bunch of reserve aircraft under a mountain in Colorado were stored just for this purpose. But the wars of the past few decades have taken a toll on all components of the system. Refineries blink out of commission; key metals are difficult to obtain; transportation infrastructure for moving commodities around on time is a disaster. The computers running it all are routinely hacked. It is like humanitarian work in Central Africa used to be: so many deaths could have been prevented by a basic medicine that you could find in any American pharmacy, and yet there was no way of attaining it. A different set of rules for what was possible seemed to apply. These days, it is getting to be the same way all over the world. The simplicity of manufacturing a jet, building an airstrip, fueling it, and supplying it with liquid sulfur compounds was one of those things that could be reliably executed at one point in time. Yet conditions have shifted; once-simple feats are today confounded by a number of different events. Hopefully China will still work this out, you think. You roll over in bed.

Your phone vibrates again. *I'm scared,* your grown child writes. From a continent away, you don't know how to answer. When they were young, a hug would fix it. Now, your survival depends on the rationality and goodwill of human beings you don't know. You think back to the slaughter of the wars of the past century, of the slide of your own country into brute falsehoods and racism and hatred of the other—history is not reassuring on this count. You try to dredge up a memory of what humans have said or thought when faced with absolute darkness, some sentiment from a great historical or philosophical or spiritual leader, but your mind is maddeningly blank.

I'm still here is the only reply you can muster. Then, another blast.

Turn to the next page.

What can we glean from this brief adventure?

Geoengineering talk often focuses on one moment—the decision to "deploy," and how or whether publics will be a part of this decision. But looking at prospective decision points muddies this notion of a discrete decision. It's also not clear exactly who these "decision makers" are. In much of our conversation about climate action, the citizen becomes a witness to history, to decision ceremonies of the powerful. Out of view are the backstories, the tiny actions that accumulated into a formal decision. It becomes hard to imagine otherwise—that geoengineering could be carried out in conversation with civil society, much less led *by* us.

Right now, geoengineering doesn't exist. Indeed, the concept is an awkward catch-all that bears little correspondence with the things it purports to describe. The UK's Royal Society laid out the term in a 2009 report, which assessed both carbon dioxide removal and solar geoengineering, also known as solar radiation management. (For a deeper understanding of how the concept of "geoengineering" came about, Oliver Morton's book *The Planet Remade* and Jack Stilgoe's book *Experiment Earth* are excellent resources.) Subsequent policy and scientific research adopted the Royal Society's framing, though it's quite possible that in the near future, the marriage of these two approaches will dissolve: a 2019 resolution brought before the United Nations Environment Assembly to assess geoengineering failed in part because it combined such different approaches. This book does consider both carbon removal and solar geoengineering, even though they are very different, because both are imagined as ways of managing an overshoot of temperature targets. Though *geoengineering* is a keyword in this book, my hope is that it is a keyword that future generations will not recognize—not because they're living it and it's become an ordinary background condition, but because it's a weird artifact of the early twenty-first-century way of seeing the human relationship with the rest of nature. This book contemplates what comes "after geoengineering" in the sense that it extends an invitation to think toward the *end goals* of geoengineering. "After geoengineering" also aims to evolve the conceptual language we use to apprehend what it means to intentionally change the climate: once "geoengineering" is a retired signifier, how do we understand these practices, and what does the new language and new understanding enable?

Even though climate engineering is mostly imaginary right now, it's a topic that's unlikely to disappear until either mitigation is pursued in earnest or the concept of geoengineering is replaced by something better; as long as climate change worsens, the specter is always there. In fact, some of the scarier scenarios result when geo-engineering isn't implemented until the impacts of climate change are even more extreme, and is therefore conducted by governments that are starting to fray and unravel. Looking at these fictional scenarios as they unfold prompts some hard questions about the optimal timing of geoengineering. Climate policy at large has been influenced by a "wait and see" attitude, where policymakers wait and see what kinds of economic damage it will cause before taking action. Research shows that even highly educated adults believe this is a reasonable approach, possibly because their mental models don't properly apprehend stocks and flows.[8] Climate change is a problem of carbon stocks, not carbon flows: the earth system is like a bathtub, filling up (an analogy used by climate modeler John Sterman and educator Linda Booth Sweeney). Reducing the flow of water into the bathtub isn't going to fix our problem unless we're actually *draining* it, too: the amount of emissions can be reduced, but green-house gas concentrations will still be rising. Wait-and-see is actually a recipe for disaster, then, because more water is flowing into the bathtub every year. Carbon removal increases the drain. It doesn't make sense to wait and see if it's needed. Moreover, it is possible that our capacity to carry out carbon removal—economically, polit-ically, and socially—could actually be greater now than it will be in a climate-stressed future.

Solar geoengineering is trickier. A wait-and-see approach makes intuitive sense: let's wait and see if society gets emissions under control in the next couple of decades, and let's wait and see if scien-tists can get better estimates of climate sensitivity and sink responses. However, there are two key limitations to note here. First, scientists anticipate that doing the research on solar geoengineering could take at least twenty years, and possibly many decades. Second, we won't know about some of these climate tipping points until we've crossed them. Imagine implementing a solar geoengineering program in order to save coastal megacities from rising seas—a plausible reason a society might try something like this. It would be desirable to do the solar geoengineering before warming reached levels where the sea level rise was locked in. But that year might only be known in hindsight, given that it's a nonlinear system. For some, this is a rationale for at least starting geoengineering research right away. A counterargument is that research is a slippery slope, and doing the research makes it more likely that solar geoengineering will be deployed.

Whatever conclusion one arrives at in this debate, the main take-away, for me, is this: There are certainly scenarios in which global society *does* figure out how to cut emissions to zero, albeit with much climate suffering (in the near future as well as our current present). Yet, if one thinks it's plausible that there won't be a significant start on this in the next decade, and that the risks of climate change are significant, it could be reasonable to look into solar geoengineering. And naturally one would want to avoid the worst-case and go for the better-case ways of doing it. There are crucial choices to be made about how it is done. For most climate engineering techniques, what is outrageous inheres not in the technology, but in the context in which it would be deployed.

Those contexts vary, but they all have two important elements. One is the counterfactual climate change scenario: How bad is climate change turning out to be, on a scale from pretty bad to catastrophic? The second is what is being done at the time to confront climate change, whether that be carbon removal, mitigation, adaptation —or nothing. These are *very* different futures, for reasons alluded to in the fictional scenarios you just navigated, and for reasons I'll go into throughout this book. The key point is that if a solar geo-engineering program is to be ended on a meaningful timescale, it will rely on mitigation and carbon removal. If a regime begins solar geoengineering, it needs to keep putting those particles up there year after year, until carbon emissions are brought down. Thus, the hard thing isn't beginning the project, but ending it: ensuring that what comes *after geoengineering* is livable. This is a battleground that's currently obscured in most discussions of geoengineering.

The definitive story of the twenty-first century, for people working to combat climate change, may be captured in one graph: the rise of greenhouse gas emissions. The line features a dramatic, tension-laden rise—and, ideally, a peak, followed by a dramatic and then gentle downslope, a resolution that accords a feeling of restoration and completion. From Shakespeare to the novel to the life course, the exposition–conflict–climax–resolution–moral story arc is a classic one. It maps nicely onto a temperature-overshoot scenario, where emissions are temporarily high but come back down. This story line lands us, the challenged yet triumphant protagonist, with 2°C of warming at century's end. These established narrative forms are how we know how to locate ourselves in an overwhelming situation; how we manage to narrate the task at hand. In these imaginaries of managing an overshoot via carbon removal, we risk simply mapping our familiar narrative form onto the problem.

As philosopher Pak-Hang Wong argues, geoengineering needs to be seen "not as a *one-off event* but as a *temporally extended process*."[9] It's not about the hero's moment of action, the climax. I would add that this re-visioning of geoengineering must be directed not just into the future, but into the past as well, thereby placing climate intervention into historical context. Future processes of both solar geoengineering and carbon removal will entail dealing with compensation or insurance for people who suffer loss and damage, working out ways to protect vulnerable people, working out who pays for it—and all that requires a reckoning with history, particularly with colonial histories of land appropriation, dispossession, and exploitation. On the international level, negotiators will have to delve into the histories of uneven development, carbon debt, and, yes, colonialism. Carbon removal can be viewed in terms of debt repayment. The addition of solar geoengineering on top of carbon removal would therefore be like living with the repo man always in the sky above you, reminding you what happens if the debt isn't paid back. Financially, we are already living in a world of debt peonage,

as Marxist geographer David Harvey points out; most of the population has future claims on their labor.[10] Now future generations are
going to have a double debt. It's not just the decision to do geoengineering that matters; it's how this carbon debt and carbon cleanup
operation is taken care of, too. The details are everything.

In reality, the resolution of this narrative curve is going to involve
struggles all along the way. The latter part of the work, the last half
of the curve towards completion, may be tougher than the first,
because decarbonizing the electricity sector by switching to solar
panels is simply easier than dealing with "hard to mitigate" sectors
or deep cultural changes, like decarbonization of aviation and industrial production, or reduction of meat consumption. Deciding to start
geoengineering is a bit like deciding to get married. It's not saying
the vows that is hard, but doing the work of the marriage. "Tying
the knot," in reality, doesn't actually mean that you're going to stay
together forever, despite the metaphor. You have to keep choosing
your spouse, or the marriage deteriorates. Solar geoengineering, in
particular, would be more like a relationship than a ceremony: and
yet much of the treatment in the literature and the press focuses on
the expensive wedding. We should instead be thinking more about
the world *after* geoengineering, because climate engineering could
be a means to very different ends.

Indeed, it has been difficult for environmentalists and the left
to engage with either carbon removal or solar geoengineering in a
forward-thinking way. Part of this is due to a fixation on the immediate need to see emissions peak—but part of it also has to do with
some serious limitations in how we think.

Copenhagen, December 2009, 1°C / 34°F
The banners unfurled under the dreary skies read "Hopenhagen."
I crossed the plaza, pigeons scattering. A historic brick building
loomed above, its rooftop scaffolding bearing the logo: "i'm loving
it." On the ground floor were a Burger King and KFC. Between
this fast-food sandwich hung a three-story advertisement sponsored

by "corporate citizens Coca-Cola and Siemens": two young, blonde boys, skinny and pale, with fists in the air, ready to heft a burden. "Earth's Bodyguards," read the caption.

I waited in the cold with hundreds of bundled-up delegates and protestors for a train to the Bella Center, where the fifteenth session of the United Nations Framework Convention on Climate Change's Conference of the Parties was taking place. We glided past a glassy office building with a several-story bright-green banner. "Stop climate change. Make COP 15 matter," it instructed us in Helvetica Light, the logo of construction corporation Skanska beneath.

At the time, climate politics seemed haunted by the specter of green capitalism. We marched under the slogan System Change Not Climate Change. While I have only a few distinct memories of this summit, they portended something quite different than our green capitalist, ecologically modernized future.

Between breaks, delegates would spill out of the conference rooms and rush to treat-laden tables in the hallways in a near melee for the best desserts. A European diplomat in a suit and a young student both reached for the last chocolate on the table, and the man in the suit slapped the confection out of the younger man's hand.

A retinue of men, dressed in suits, swept briskly through the corridor like a cold wind. The man in the center was the focal point; the rest flanked him, like a military formation. I flattened myself into the side of the hallway as they passed. It's an unremarkable thing, people passing each other in a nondescript corridor, but I felt chilled. "Did you see Robert Mugabe? He's here," someone whispered to me a few minutes later.

A tent, in the rain, in the "free city" of Christiania. I listened to Naomi Klein and other activists muster the forces. We drank mulled wine to keep warm and waited for the police to sweep in with their water cannons and tear gas; there was a rumor that they were coming. (They came.)

There was a kind of power that crackled in the air. Every time it manifested, it surprised me. I was expecting a climate summit to be a rather stuffy and formal affair, filled with acronyms and technical jargon. The injunctions of green capitalism postered around the city seemed pleading, thin, compared to these older and more primal forms of power. Hugo Chávez, speaking at the summit, said that "a ghost is stalking the streets of Copenhagen ... it's capitalism, capitalism is that ghost." Chavez declared, "When these capitalist gods of carbon burp and belch their dangerous emissions, it's we, the lesser mortals of the developing sphere who gasp and sink and eventually die." I can understand the sentiment—particularly when it comes to the unevenness of climate impacts and the brutality of the historical record. As ecological Marxist theory argues, capital accumulation and the treadmill of production is a central factor in global environmental degradation—a thesis I'm onboard with. Nevertheless, I don't think that *green* capitalism was the ghost roaming those halls. Perhaps we were focusing on the wrong ghost.

Those of us schooled in keeping watch against green capitalism would naturally read geoengineering as capitalism's next move in self-preservation. I'm skeptical of this, because I don't see the evidence that capitalism is capable of acting in its own long-term benefit—especially not consciously, on the scale and temporality of mobilization that this intervention would require. (Although oil companies might be: a slightly different prospect that is discussed in Chapter 8.) But capital is something of a headless monster, incapable of this kind of macro-level, strategic, long-term thinking. In the face of what could be an existential crisis, innovation is flowing toward hookup apps and making sure porny advertising doesn't get stationed next to famous brands. This is where capital's attention and money is directed; as anthropologist David Graeber observes, technological progress since the 1970s has been largely in information technologies, technologies of simulation. Graeber notes that there was a shift from "investment in technologies associated with the possibility of alternative futures to investment in technologies that furthered labor discipline and social control"[11]—in other words, it's a big mistake to assume capitalism is naturally technologically progressive. In fact, he suggests, "invention and true innovation will not happen within the framework of contemporary corporate capitalism—or, most likely any form of capitalism at all."[12] I agree—we've seen numerous terrific ideas since the 1970s in alternative energy, and even in carbon removal, but they've been constantly thwarted or shelved. Whatever form of capitalism we're living in now, it doesn't seem like a system in which carbon removal is going to evolve. The derivation of capitalism we're coping with is predatory, inelegant, and fragmented, seemingly incapable of delivering fixed-capital tools like carbon capture and storage or transformative bioenergy to extend its lifespan.

Critical theorist McKenzie Wark asks: "We think within a metaphysical construct in which capital has some eternal inner essence, and only its forms of appearance ever change ... But what if the whole of capitalism had mutated into something else?"[13] Wark speculates on the emergence of what he calls the "vectoralist" class, a new postcapitalist ruling class that owns and controls the means of producing information: the vectors. This is actually worse than capitalism, Wark argues, because the information vector can render everything on the planet a resource.

So what does all this mean for geoengineering? If capitalism is focused on vectoral control and ineffective when it comes to ensuring the material conditions of its own existence, solar geoengineering would be done by states or not at all. As for carbon removal, the question is this: If zombified neoliberal capitalism isn't going to build up CCS and carbon removal in order to save itself from planetary crisis, who's going to do it?

We, the workers and voters, will have to decide to force the removal of carbon from the atmosphere. And we should—those of us living in the global North, in particular. A whole host of commonly accepted moral principles align with carbon removal: "clean up your own mess," "the polluter pays," the "precautionary principle," and others. Moreover, doing carbon removal in a socially just and environmentally rigorous manner is not just morally desirable —*it is actually a precondition for emissions going net negative.*

There are basically two levels to carbon removal, as I see it. Level 1 involves niche, boutique, aesthetic, or symbolic removals. This is the biochar at your farmer's market, the wool beanie grown with regeneratively grazed sheep, the shoes made with recycled carbon, water carbonated by Coca-Cola with carbon captured directly from the air. It is cool. Advocates see it as the first step toward reaching Level 2. You don't want to knock its fragile emergence, because it's important for generating momentum and raising awareness of carbon removal. But it's geophysically impossible that it will "solve" climate change.

Level 2 is the gigaton-scale removals that could actually lower greenhouse gas concentrations. Call it "climate significant." It's waste cleanup; pollution disposal.

How does one get from Level 1 to Level 2? Some people think it will naturally happen, just as cleantech—renewable energy— "naturally" becomes cheaper and scales. But unlike cleantech, Level 2 is a cleanup operation; in general, these scales of storage and disposal don't generate usable products. I asked Noah Deich, executive director of the nonprofit Carbon180, about these middle-range pathways from demonstration to disposal scales, because his organization has done significant work articulating policy proposals for carbon removal. In the near term, Deich sees a threefold approach, or a "stool with three legs." One is moonshot research and development across the technology and land sectors. The second is supporting entrepreneurs to bring promising ideas to market. Lastly, he notes, "we need to change policy so that there's sufficient funding for the research and development, but there are also large-scale markets, so that those entrepreneurs and those land managers can access those markets at a meaningful scale." The near-term actions he identifies include engagement of universities in research and development, starting up an incubator for carbon tech, and policy work such as implementation of tax credits for CCS and the inclusion of carbon farming in the US farm bill.

When I remarked that the middle time frame seemed fuzzy, Deich replied, "The middle part will remain fuzzy, because I think it's iterative." You get started with technology in existing markets, which creates jobs and investment opportunities, he says. Success begets policy support, whether it be government or corporate, which begets more markets, and it becomes a reinforcing cycle that snowballs. "If we're able to create incentives for taking that carbon out of the air, I think it's reasonable that we'll be able to ratchet up those incentives and build that broad political coalition that's both durable and meaningful to do this at large scale."

Yet I am less and less convinced that there is a clear route from Level 1 to Level 2, nor that the first would naturally progress to the next. Level 1 is what our current set of policies and incentives can accomplish, with a lot of work from think tanks, NGOs, philanthropists, and the like. Level 2 requires a massive transformation: economic, political, cultural. It implies that we decide to treat carbon dioxide as a waste product and dedicate a significant portion of GDP to cleaning it up, at the least. It would require profound state regulation—the same sort that's needed for strong mitigation, and then some.

There is sometimes a hope among environmentalists and social justice advocates that confronting climate change will itself bring about social transformation—that it will flip us into a new narrative that could take on the climate pollution challenge. As cultural theorist Claire Colebrook writes,

> From Naomi Klein's claim that climate change is the opportunity finally to triumph over capitalism, to the environmental humanities movement that spurns decades of "textualist" theory in order to regain nature and life, to wise geo-engineers who operate from the imperative that if we are to survive we must act immediately and unilaterally, the end of man has generated a thousand tiny industries of new dawns.[14]

However, I think there are plenty of scenarios where we deal with climate change in a middling way that preserves the existing unequal arrangements, leaving us not with a new dawn, but with a long and torturous afternoon. Replacing our current liquid fuels with synthetic, lower-carbon fuels produced with direct air capture and enhanced oil recovery would be one version. But those dawn-less scenarios are not necessarily geoengineering scenarios, and vice versa. There are both horrifying and mildly likeable scenarios for how carbon removal might be accomplished. The horrifying ones are easy to conjure to mind, while the likable ones stretch the imagination. It would be easy to tag best-case carbon removal scenarios as utopias—even though they would actually be worlds that have failed to mitigate in time, representing at best a muddling through. That's where we're at: even muddling through looks like an amazing social feat, an orchestration so elaborate and requiring so much luck that people may find it a fantastic, utopian dream.

We can maximize our chances of muddling through by engaging proactively with both carbon removal and solar geoengineering. However, binary thinking about climate engineering has made it difficult for progressives to create a dialogue about how engaging with these emerging approaches might be done. Climate engineering has been stuck in the realm of "technology," rather than understood as a variety of practices that include people in various relationships with nature and each other. To free ourselves of these binaries and imagine a different kind of strategy-led engagement, it's valuable to articulate a best-case scenario for how these practices could unfold.

Rigid binaries leave climate engineering in the hands of elites

There's an abyss in contemporary thinking about the role of industrial technology in coping with climate change.

On one side of this abyss are people who appraise the potential of technology optimistically, but fail to articulate any real historical awareness of how technology has developed in and through contexts that are often exploitative, unequal, and even violent.

On the other side of the abyss are thinkers who, on the contrary, have a deep understanding of colonialism, imperialism, and the historical evolution of capitalism, but dismiss technology as a useful part of responding to climate change.

This cleavage leaves little room for critical discussion of how technologies might be used to further climate justice. It makes it impossible to imagine, for example, democratically controlled industrial technology that doesn't exist to "conquer" nature. Today, most left thinking has abandoned the "streak of admiration for the productive forces as the instruments of a conquest of nature that will ultimately usher in communist affluence for everyone," as human ecologist Andreas Malm has observed.[15] But this abandonment did not immediately lead to a coherent articulation of a view of technology that is collective or cooperative, or that works with nature.

I am not the first to observe this. A number of calls have emerged recently for the left to think differently about industrial technology. Geographer Matthew Huber, for one, suggests that "Marx believed that there is something inherently emancipatory about large-scale industrialization, and ecosocialists need not be so quick to dismiss this possibility." He asks, "What if the phrase 'development of the productive forces' was not simply equated with the expansion of dirty industrial production based on coal, oil, and gas and instead represented the full development of industrial energy systems based on cleaner and renewable fuels?"[16] Sociologist Jesse Goldstein, in *Planetary Improvement*, his critical ethnographic analysis of cleantech, observes that "the sociotechnical capacity is out there to transform the world in any number of ways," but realizing emancipatory visions will require "killing the investor" in our minds, "thereby liberating our imaginations, our sciences, and our technologies from the narrowing logic of capital."[17]

Others calls to embrace industrial technology are more strident. Science journalist Leigh Phillips, in his book *Austerity Ecology and the Collapse-Porn Addicts: A Defence of Growth, Progress, Industry, and Stuff*, derides the left's small-is-beautiful, local-retreat tendencies, pointing out that the left was not always this way:

> Historically, when we criticised the failings of the market, the left had no particular quarrel with industry, let alone science, technology, or medicine. We celebrated modernity. Rather, our demand had always been that the fruits of civilization should be extended to *all* of humanity. When did we turn away from the idea that capitalism was the problem, and begin to believe that it was modernity instead, or even the advent of mankind itself, that was the problem?[18]

Phillips's book attempts to answer that question by arguing that austerity ecology is an incarnation of a "very old, dark and Malthusian set of ideas that the left historically did battle with." Phillips sees austerity and degrowth as "mathematically and socially identical." To solve the global biocrisis, more is needed: more growth, progress, industry, and civilization. He asserts that "it will require significant ingenuity to engineer a reverse of the processes we have inadvertently set in motion, likely even some way to produce a carbon-negative economy for a period," with hundreds of innovations that will come from the most advanced research laboratories and factories. "By turning its back on the possibility of such technologies, on the very idea of progress, green anti-modernism *actually commits us to catastrophic climate change.*" The conclusion is that retreat is not an option; we must accelerate modernity. In some kind of company with this book is Nick Srnicek and Alex Williams's *Inventing the Future: Postcapitalism and a World without Work*. Srnicek and Williams reject the localism of "folk politics" and call for repurposing technologies, pointing out that it's not just about seizing the means of production, but inventing *new* means of production.[19]

Xenofeminism, as articulated in the *Xenofeminist Manifesto* by the collective Laboria Cuboniks, also seeks to "strategically deploy existing technologies to re-engineer the world." They ask, "Why is there so little explicit, organized effort to repurpose technologies for progressive gender political ends?" The real emancipatory potential of tech, they claim, is unrealized:

> The excess of modesty in feminist agendas of recent decades is not proportionate to the monstrous complexity of our reality, a reality cross-hatched with fibre-optic cables, radio and microwaves, oil and gas pipelines, aerial and shipping routes, and the unrelenting, simultaneous execution of millions of communication protocols with every passing millisecond. Systematic thinking and structural analysis have largely fallen by the wayside in favour of admirable, but insufficient struggles, bound to fixed localities and fragmented insurrections.[20]

Again, there is a call for a wider scale of analysis and practice than local efforts or folk politics can allow for. The *Xenofeminist Manifesto* further asserts that "suggestions to pull the lever on the emergency brake of embedded velocities, the call to slow down and scale back, is a possibility available only to the few—a violent particularity of exclusivity—ultimately entailing catastrophe for the many."

While these currents are swirling on the radical left, they haven't seeped into the environmental mainstream. There, the conception of an industrial technology that works with nature is limited to solar panels and wind turbines (as long as they are not in anyone's backyard). Otherwise, industrial technology is still seen as that of the dark satanic mills—and certainly, there are plenty of extractive operations around the globe that reinforce this view. Technology and capitalism remain conflated (and the heavy government subsidies received by many transformative technologies are elided from view). So when it comes to geoengineering, many environmentalists have adopted a simple refrain: "We don't need geoengineering, we need *x*." This is a familiar formula, where *x* may be sustainable, ecological agriculture. Or system transformation. Or degrowth. Geoengineering serves as a foil for the beautiful *x*, the blossoming future we *really* want.

Let's look more closely around how these binaries are articulated, and where they originate. One formulation of the binary is to view geoengineering as conflicting with real, transformative change. This is a compelling structure of thought because it focuses on changes that advocates *actually want*. The ETC Group, a civil society organization critical of the ways many emerging technologies are conceived and wielded, warns against geoengineering as a "perfect excuse": "Geoengineering offers governments an option other than reducing emissions and protecting biodiversity."[21] Journalist and activist Naomi Klein declares that "the fact that geoengineering is being treated so seriously should underline the urgent need for a real plan A—one based on emission reduction, however economically radical it must be."[22]

Advocates of geoengineering research, on the other hand, tend to believe that mitigation will not be enough to stave off the worst of climate change. One counterargument to the no-plan-but-mitigation position is posed by Harvard solar geoengineering researchers David Keith and Josh Horton:

> At the risk of oversimplification, this line of argument essentially involves rich-country commentators criticizing solar geoengineering in an effort to shore up mitigation as their priority domestic climate policy, while ignoring the potentially huge distributional advantages SRM [solar radiation management] might confer on the world's poorest in the global South. Their deeper motives vary, from a sense of moral indignation over shirking (Hamilton) to neo-Luddism (ETC Group) to anti-corporate ideology (Klein) and beyond. Whatever the reasons, the resulting admonition not to research SRM for fear of its policy implications for industrialized countries, at the expense of possibly enormous welfare gains in developing countries, is ethically disturbing in a global moral context.[23]

Retorts to *this* counterargument have included assertions that research advocates are being insincere about their concerns for vulnerable people, or limited in their thinking. Environmental policy scholars Jane Flegal and Aarti Gupta, for example, point out that the "performative power and political implications of specific expert visions of equity, evoked as a rationale to undertake solar geoengineering research, require continued scrutiny"—in part because these expert visions are filtered through a technological frame where equity is understood only narrowly.[24] Moreover, philosopher and activist Kyle Whyte observes that "the argument that geoengineering is actually intended to sort of save or support indigenous people is actually hard to maintain because it's not temperature rise or unpredictable precipitation that are really the problems."[25] Rather, colonialism and global inequality are at the heart of climate change— so if you care about the poorest in the global South, shouldn't those be first and foremost in the discussion?

I agree wholeheartedly with the critiques offered by Whyte, Flegal, Gupta, and others of the solar geoengineering research enterprise. Yet, to me, it is presumptuous to entirely cross off an idea that could, in a future scenario with runaway climate change, alleviate much suffering in places with less capacity to adapt to changing conditions. Systemic change is absolutely necessary. But geoengineering does not have to substitute for transformative change —in fact, to work well, geoengineering *requires* systemic change, because *responsible solar geoengineering requires carbon removal, which requires renewable energy.* This scale-up of renewables and carbon removal is only accomplishable with massive social and political transformation. The best-case solar geoengineering scenario is only achievable with dramatic social change. At the same time, critics rightfully worry that fossil fuel actors and other elites will use solar geoengineering to forestall social transformation, and the same concern applies to carbon removal. Yet at this point in time, a blanket rejection of carbon removal, in particular, comes off as an aesthetic luxury.

At the roots of this binary between geoengineering and social transformation, I think, are contemporary struggles around agriculture and food sovereignty. Contemporary agriculture is riddled with binary constructions: there's holistic agroecology versus reductionist and mechanistic industrial agriculture; resilient agroecology versus techno-fix drought-resistant crop breeding. Geoengineering gets mapped onto these preexisting binaries, leading to the impression that geoengineering represents one distinct set of options, and agroecological transformation another. It is the mindsets, values, cultures, and systems behind these practices that agroecology advocates perceive to be in conflict. "We need regenerative farming, not geoengineering," argues teacher and activist Charles Eisenstein in a 2015 *Guardian* op-ed.[26] The ETC Group, for its part, opposes geoengineering as a false solution to climate change, along with proprietary climate-ready crops, and supports peasant-led agroecological responses to the climate crisis.[27] Environmental NGO Biofuelwatch calls climate engineering techniques like bioenergy with CCS a "distraction of attention away from genuine and credible ways of sequestering carbon: agroecology and ecosystem regeneration."[28]

One more binary merits discussion here—one that lies *within* carbon removal practices themselves. Scientific and policy reports categorize these practices as *biological* (or natural) climate solutions, on one hand, and *engineered* solutions, on the other. This is reasonable to some degree: carbon can be sequestered either by growing more things (the focus of Part I of this book), or by burying it geologically (discussed in Part II of this book). In climate policy discourse, though, the divide between the industrial and the biological is playing out within carbon removal itself, with one side often (but not always) privileged over another. At the brink of desperation, activist groups like the Sunrise Movement, The Climate Mobilization and Extinction Rebellion are beginning to engage with demands for carbon drawdown. But many climate justice groups are ambiguous about how the drawdown can be achieved, and there is often a cognitive gap between the demand for drawdown and the scale of industrial activity required to accomplish it.

These overlapping binaries—geoengineering versus real change, geoengineering versus agroecology—obscure the reality that there is a spectrum of ways of doing, enacting, practicing, deploying, or implementing climate intervention. The implementation does not inhere in the technology. Sticking rigidly to these binaries keeps us from seeing possible futures: it gives the terrain for shaping climate engineering over to the few.

Climate intervention as practice

Climate engineering is not a monolithic "technology," but a variety of *practices* (or activities), and actors have some level of choice about *how* they will practice it. Jack Stilgoe, in *Experiment Earth*, suggests that we need to view geoengineering not as a noun but as a verb: "Viewed as a set of technologies, geo-engineering resembles no more than a mixed bag of half-baked schemes. If we take literally the meaning of 'geoengineering' as a present participle, it becomes a project, a work-in-progress." Stilgoe suggests that viewed this way, geoengineering is a form of governance.[29] Yet often, climate engineering is still viewed as a "thing," an artifact that comes out of this box of emerging technologies, alongside genetics, robotics, information technology, and nanotechnology. In some very important ways, climate geoengineering is not like these other emerging technologies.

Emerging technologies are sometimes imagined to spur new long waves of innovation, or "Kondratiev waves," after the economist who theorized them. First there were textiles during the Industrial Revolution, followed by steam and rail, and then electricity, and then automobiles and the petroleum age, and later information technology; some new transformative innovation comes along every forty to sixty years. This theory isn't a particularly fashionable one in mainstream economics right now, but I mention it here because it helps illustrate an interesting point: geoengineering is never going to be like those other emerging technologies, all of which played transformative roles in economies. Instead, carbon removal is likely to be analogous to waste control: a massive industry, but not a transformative one. Solar geoengineering, in particular, is ameliorative and not generative; that is, it doesn't generate new wealth. Its growth is necessarily limited, and the number of actors that would be needed for its realization is also limited. What would be their motivation to engage? Researchers can gain individual glory and social capital, maybe, but it is difficult to imagine solar geoengineering being accomplished through an investment/shareholder corporate model. When it comes to real-world benefits, solar geoengineering is so broad, crude, risky, and low profit that it is best viewed as a global public project. The benefits of carbon removal would also be a global public good, though potentially one with damaging effects in particular places, depending on policy design. In short, it's easier to envisage climate engineering as undertaken for the benefit of the many.

The point here is that solar geoengineering is not actually "a technology"—indeed, most of these socio-technical systems aren't. The planes and nozzles, and the software that drives and creates solar geoengineering would be technologies, as are the computer models that indicate it would cool the planet. But while solar geoengineering *relies* on such technology, it would be more than that. When we put it in the "technology" box, it becomes the domain of technology experts, and we fail to see what else it is; the social life of the intervention is obscured.

If geoengineering is not simply an emerging technology, what is it? Let's consider three alternative frames: development interventions, humanitarian interventions, and infrastructure. None of these is sufficient on its own to understand geoengineering, but each of them can illuminate something about what geoengineering could be.

Development-speak is filled with the language of intervention, from community-level health or nutrition interventions to macro-level economic interventions. Climate-related development interventions have included things like education in climate-smart agricultural practices for farmers, or the institution of clean cookstoves. Interventions like these are usually designed to have multiple social and climate benefits. These programs are also monitored and evaluated, an aspect that makes them a useful analog to solar geoengineering. For solar geoengineering, too, would require implementation and management across years and years, and continual monitoring and evaluation. When it comes to carbon removal, there are some ways in which it is already conceived of as a development intervention—something that is apparent, for instance, in community forestry or biochar projects.

"Intervention" also comes up in humanitarian work, where it implies intervening in some disastrous situation, either through military force or humanitarian aid and relief, and often in an international or nongovernment partnership. There's generally no direct profit motive, although private contractors do profit from humanitarian work. Humanitarian interventions constitute a relevant parallel because of the emergency rationale; like for solar geoengineering, the interventions aim to "save" or "stabilize" something. Humanitarian interventions tend to borrow heavily from military language, describing their projects in terms of missions or deployment. The same is true for the language typically used to speak about solar geoengineering.

Much of the earliest work on geoengineering used the term "climate intervention." Take, for instance, the 2010 "Asilomar International Conference on Climate Intervention Technologies"; the 2015 National Academies reports on climate intervention also adopted this term. It is used synonymously with both "climate engineering" and "geoengineering." So why isn't "intervention" a go-to frame for understanding climate engineering? One reason is that intervention in the development or humanitarian context focuses on the action; but with geoengineering, focusing on the action seems "premature." Researchers are careful to specify that right now we're only talking about research, and not thinking about deployment. There's a carefully constructed gap: On one side is an idealized world where we can run models, and where solar geoengineering is abstract, and therefore safe. On the other side is the world of imagined deployment, which leads one down the path of imagining *particular* deployment scenarios; using analogies like intervention pushes the conversation from research talk into deployment talk. Another reason not to employ these analogues concerns optics: the "humanitarian intervention" comparison sounds like greenwashing, an Orwellian euphemism. I don't necessarily find these terms euphemistic, though, since neither the term "development" nor "humanitarian" connotes *goodness* to me. Rather, they refer to specific goals and projects. The development project has worked out quite terribly in many places, extending Western colonialism, trapping poor nations in debt, and transforming communal social relations into exploitative ones. Similarly, when it comes to humanitarian intervention, we've seen plenty of disastrous results. The fact that many of these social interventions have gone so terribly wrong, in fact, is precisely the reason why it is important to think about geoengineering with reference to these examples.

While part of climate engineering resembles a programmatic intervention—something constructed, through time, with specific goals and management superstructure—another part of is more like infrastructure—fixed, heavy, material. It is natural to talk about infrastructure when considering carbon dioxide removal, since industrial forms of carbon removal require large-scale pipelines and facilities. Similarly, reforestation and soil carbon sequestration can be seen through the lens of "ecological infrastructure." Solar geoengineering, by contrast, appears more intangible, ethereal. But its infrastructure is simply flexible, and approaches like stratospheric aerosols are reliant on existing fixed infrastructure such as runways, factories, and mines. The moment of infrastructural development is a flash point for contestation; the concrete image, whether it be a diagram or architectural rendering, makes it real enough to fight against. Until the infrastructure is imagined, we're still in the sci-fi fantasy space of floating cities. It's also helpful to think of geo-engineering as infrastructure, as environmental humanities scholar Anne Pasek points out in her in-depth analysis, because doing so evokes the care and maintenance required.[30]

By thinking of geoengineering as infrastructure, we position ourselves to heed the lessons of past megaprojects. The most familiar megaprojects are multibillion-dollar infrastructure projects: the Channel Tunnel, the Øresund Bridge, the Three Gorges Dam, the Hong Kong International Airport, and so on. Megaprojects can involve infrastructure (dams, ports, and railroads); extraction (minerals, fossil fuels); production (fighter aircraft, chemical plants, and manufacturing parks); and consumption (tourist installations, malls, and theme parks), notes Bent Flyvbjerg, a megaprojects expert. He writes that megaprojects are part of a remarkably coherent story, what sociologist Zygmunt Bauman has called the "Great War of Independence from Space." They imply mobility, liberation. He talks about the end of geography, the death of distance, and so on.[31] Perhaps you remember this zeitgeist from the early 2000s, when the internet was new and transformative, before we knew it would give us so many cat videos and listicles and trolls. When "globalization" was still a buzzword, before the financial crisis and the failed interventions in Iraq and Afghanistan. Critics of geoengineering tend to locate the psychological roots of climate engineering in postwar, big science techno-optimism, in 1950s thinking. But it is equally useful to regard it as a phenomenon born of the early 2000s, a more globalist moment.

The paradox of megaprojects, Flyvbjerg writes, is that even as more and larger infrastructure projects are proposed and built, they evidence strikingly poor performance records in terms of economy, environment, and public support. Geographers Ben

Marsh and Janet Jones point out, however, that when you take into account symbolism, this is only an *apparent* paradox—economic performance is not the only measure of success, as megaprojects are planned and executed for a symbolic value that can be more stunning than their fiscal value.[32] Infrastructure inscribes cultural messages in the landscape; it expresses both authorship and authority. Mega-engineering projects are hyperlegible; scale becomes a design factor. And so, Marsh and Jones observe, while power is the "foremost statement" of large landscape projects, the actual messages are diverse, ranging from abundance (the Hanging Gardens of Babylon) to security (levees on floodplains). But if infrastructure is about messaging, it begs the question: Could climate engineering projects be more effective as a symbolic strategy than as a material one?

The shortcomings of large infrastructure projects have generated suspicion about megaprojects, suspicion which may be transferred to solar geoengineering. Flyvbjerg points out that the documents of megaproject preparation—cost–benefit analyses, financial analyses, impact statements—are called into question and denounced more often than analyses in any other professional field. It's common to have cost overruns of 50 to 100 percent or more, and demand forecasts that are wrong by 20 to 70 percent.[33] As Flyvbjerg writes, the key problem is lack of accountability, not lack of technical skills or data. Forecasts are manipulated, special interest groups promote projects at no cost or risk to themselves, contractors underestimate costs and risks—meaning the real costs and risks don't surface until construction is underway. This happens drastically in defense contracts, for example, with taxpayers footing the bill. "Appraisal optimism" is a generous way to put all of this, and the collective experience with megaprojects is a cautionary tale for climate engineering.

But solar geoengineering has another perceived relationship to infrastructure: rightly or wrongly, it is seen as a blanket infrastructure *preserver*. There are quite reasonable concerns that solar geoengineering is a way to avoid changing this other $13 trillion infrastructure of fossil fuels, implying a workaround for the phaseout of new fossil fuel plants "prematurely," and saving assets from being stranded. But infrastructure always changes; as science historian David Nye writes, "Even the largest and most successful technological systems eventually lose momentum."[34] He offers examples from the coal distribution infrastructure of the last century: abandoned coal yards, wagons, bins, and chutes. Think of the rapid build-out of communications infrastructure, or of the structures left behind in the wake of the exodus from agriculture. History shows infrastructure's impermanence and offers lessons on how people in particular places react to the changes.

Understanding geoengineering as a program, practice, project, intervention, infrastructure, and so on might make the concept seem sprawling. But consider what environmental scientist Brad Allenby writes about jet technology: a traditional life cycle analysis counts the use phase of the jet itself, but really, the jet enabled a global tourism industry to spring up, which, as Allenby points out, has probably had more impact on the biosphere than anything since European colonialism—in terms of knitting together population centers but also creating new disease and invasive species vectors. "So, should these profoundly systemic effects be considered as one contemplates design of a jet aircraft?" Allenby asks.[35] The complexity becomes staggering. Nevertheless, a systemic perspective is necessary. We need to understand the indirect effects of interventions, the parts of the system that behave differently at different scales, and so on. We don't currently have the institutions, training, and methods to adequately look at something like climate engineering from a systemic perspective. Even in the halls of the world's most vaunted universities, the discussion and framing of both solar geoengineering and carbon removal is extraordinarily thin, stripping out the social complexity.

Looking at the history of megaprojects and failed interventions, the question looms: How do we prevent failed attempts at geoengineering? Solar geoengineering isn't a regular project: the damage of an aborted or poorly executed project isn't measured in costs or missed opportunities to invest in other things, but in ecological dangers. In fact, the fiscal cost is the least significant aspect of the failure. The worst-case scenario here might not even be extreme climate change, or that solar geoengineering is done, but that solar geoengineering is attempted poorly.

In short, rather than simply being emerging technologies, both solar geoengineering and carbon removal would be practices that have aspects of infrastructure and social intervention. They must be wrested from the realm of technology—where only experts are permitted—and seen through the prism of projects, programs, and practices if civil society is going to attempt to shape them in a meaningfully democratic way.

What would a better-case geoengineering look like?

Swift and deep decarbonization is the best-case climate future. But again, the specter rears its head that this won't happen in time to avoid extreme climate impacts. There's a genuine possibility that only clear and significant climate impacts will motivate real action, and by then, a significant level of warming will be locked in and looming.

Many forms of geoengineering may be dangerous or unworkable. Even the basic idea of drawing down greenhouse gas concentrations has its unknowns. When carbon dioxide is emitted, about a quarter goes into the oceans, about a quarter of it is stored in ecosystems, and about half of it remains in the atmosphere. So removing 100 Gt from the earth system would roughly mean reducing the total amount in the atmosphere by only 50 Gt. If carbon removal reduced carbon concentrations in the atmosphere, CO_2 in the oceans would gradually transfer back to the atmosphere, a flow that is complicated to model because of how the layers of the ocean mix. Carbon cycle feedbacks could reduce the effectiveness of carbon removal, or perhaps enhance it.

Yet the possibility of climate catastrophe makes thinking through the best-case use of all these approaches a valuable thought experiment. For if their best-case use, under close examination, is unattainable, perhaps the idea had better be removed from discussion —which may not be an easy feat.

Is there a synthesis to be had between geoengineering and sustainable agriculture (and earth care)—a better "geoengineering"? Or, perhaps not a collapsing or synthesis of these two practices, but a new term and framework of understanding to be created? Indeed, "climate restoration" advocates and other groups may create the ground for this. This book profiles a handful of people who are articulating visions that transcend this binary. In the following chapters, we'll explore possible contours of a world during and after geoengineering, through the voices of some of the scientists, entrepreneurs, and activists that could have a role in shaping this world.

Here's an after-geoengineering test for geoengineering proposals: Is this proposed program or project likely to produce a livable world 200 years from now? By making a best-case scenario vivid, it becomes easier to grasp the magnitude of the challenge, and to see how it runs through many aspects of everyday life, in ways viewing geoengineering simply as a "technology" may not.

So often, climate futures are described in terms of mathematical pathways or scenarios, behind which are traditions of men gaming out possible futures. In this book, I've used fiction to do what nonfiction writing cannot do as well: to make the future less empty, to populate it with embodied lives and emotions. For it is *people*, with bodies and lives, who have to experience climate change; climate futures aren't just about geopolitical and temperature events. While there are many reasons for climate inaction, one of them is that climate change has been coded as an issue of "science" or "politics," of serious and hard stuff, with the human content and emotions separated out. This division performs, yet again, the master nature/culture binary that this book seeks to do away with. Fiction is one way to bring back in that which has been parceled out of the climate change conversation; this book's hybrid form is meant to create a synthesis, to bridge that binary. Although such hybrids can seem like strange creatures, the aim is to invite the reader into alternative imaginations of the future, for small fractions of this book, and to experience something different.

This book is about the future—including scenes that speculate about the end of the century—but that future starts right now. The actions we take during the next decade will drastically shape what kind of world our descendants dwell in.

Part I: Cultivation

1

Cultivating Energy

What has drawn the Modern World into being is a strange, almost occult yearning for the future.

—Wendell Berry

La Jolla, California, July 2014, pleasant weather

Unlike a pistol, designed for fluidity of motion, the helium-powered gene gun was awkward to hold. The lab group circled nervously, wearing bulky pilot-grade noise-canceling headphones to protect us from the sonic force. Glasses: check. Lab coats: check. Gloves: check. All this protected us from the tool and the microscopic amounts of liquid we were going to use it on. We handed the gun (manufacture: Bio-Rad) around the circle ritualistically as we took turns modifying life.

The gun was loaded with particles of gold that we had coated with DNA. We were shooting the DNA-laden gold onto petri dishes prepped with algae cells. The microscopic gold particles can blast through the cell walls of the algae, allowing the DNA to get in. It is an inelegant technique, not very efficient, but enough of the particles make it through. The point was to make the algae colorless. It was a training exercise, a drill. We were enrolled in a course that taught the basic recipes of manipulating life—not as part of a college degree, though we were on a university campus in Southern California. Rather, it was a part of an effort to train new workers in the biofuel industry. Green jobs. You didn't actually need to know much of the scientific grounding of microbiology to follow the recipes. They worked anyway.

Blasting the algae with the gold particles only took a few moments. Afterward, I took off the heavy headphones and stepped out of the lab into the light and sound of the day: the eucalyptus trees rustling, the blue sky, the uncanny thunder of the jets blasting away from the nearby Miramar air base, doing their laps across the skies above the biotech hub of San Diego. In the manicured office parks of La Jolla, companies are seeking to turn algae into value: protein, medicine, and fuel. It is this latter transformation, algae into fuel, that intersects our story of climate intervention. Of all the crops that can be cultivated to suck up carbon, algae is one of the most promising.

Cultivation is one of the most tried-and-tested techniques to change the carbon balance. Plants, of course, take up carbon, but they can also be used for storing energy. Plants can also be made into physical end products, which is the basis of the bioeconomy: an economy based not upon dead matter, but upon living things. The bioeconomy probably "exists" more than "geoengineering" does, though it still dwells in the imaginative realm of charismatic meta-categories. The promise of the bioeconomy is that it allows throwaway, mineral-based goods to be replaced by renewable ones (which is why it is sometimes called the "circular bioeconomy," as products aren't thrown away, but return to the circle of life to be processed into new ones). The bioeconomy is also often defined in terms of products or sectors that compose it: food, agriculture, paper and pulp, forestry and the wood industry, fisheries and aquaculture, bio-based industries, biochemicals and plastics, and enzymes and biofuels. It's an intuitively attractive concept, given how biophilic humans are.

Engineering new uses for plants and setting up a cultivation-based economy sounds green-futuristic, but it's actually an old dream. The bioeconomy vision had a precursor almost a century ago, during the dawning of the petroleum age in the 1930s and 1940s. In the United States, widespread production of oil emerged against the backdrop of the Great Depression, rural out-migration, and a farm production surplus (not to mention a looming war). Could there be a comprehensive solution to the social and economic ills of the time? You might be thinking of the New Deal, but industrialists like Henry Ford and scientists like George Washington Carver had another idea: chemurgy. At its simplest, "chemurgy" meant using farm products for stuff besides food. With the goal of productively using agricultural waste, chemurgy had a strong ethos of efficiency, as well as of self-sufficiency. To some of its promoters, using the wealth in plant resources also offered the dream of universal abundance and a peaceful world.

Vegetable or mineral? Living fuel or dead fuel? In the chemurgy movement, we can see traces of the division between biological and geological systems, a binary that lingers in the carbon removal conversation today. Leading figures of the time knew that petroleum was not going to work out well for us. Petroleum reserves were imagined sufficient to last only a matter of decades. As journalist Christy Borth wrote in 1942, "We do not have to go into the bowels of the earth for fossilized sunshine. We need not destroy one another with warfare because some nations happen to have fossilized sunlight and other nations have none." The necessary shift, as Borth put it, was from dependence upon "the sub-surface *fixed* resources to the surface *flow* resources."[1] Chemist William Hale, the husband of Helen Dow of the Dow Chemical family, also saw the crisis facing the world as part of a transition from the machine age to the chemical age. He also saw the necessity of developing "agricrude" fuel, or ethanol. It seemed nonsensical to burn fossil fuels when renewable fuels were so easily obtainable, and Hale lamented that the "selfish interests" of the petroleum industry had "hoodwinked" the American people. The cracking of petroleum for gasoline was "one of the most wasteful deeds of man," according to Hale, especially since future generations might need that petroleum for their endeavors. "To burn petroleum with the reckless abandon of today is to withhold from posterity an asset in which they, as much as we, have equal rights." Hale's words are haunting:

After all, these storehouses of gas, petroleum, and coal are precious endowments to man by nature. Human decency should teach us not to destroy them indiscriminately. Human kindness should teach us to preserve as much as possible for our children. Nevertheless, utter profligacy has gained the upper hand; as pirates and plunderers we seem destined to go down the road of defeated nations.[2]

Why has nobody heard of chemurgy? It was a startup movement; a risky field. It achieved some successes: four new national research laboratories, the beginnings of an American flax paper industry, the development of the Southern pine industry, and wood-to-energy efforts all owe their dues to the field. But the word had largely fallen out of favor by 1950 (and one wonders if "geoengineering" will be similarly obsolete in a few decades). The simplest explanation for chemurgy's failure is that petroleum was a cheaper material for fuel and other goods. But there were other factors underlying chemurgy's failure to take off. There was often a wide gulf in the scale of chemurgical pilot projects or prototypes and full-scale production. Some analysts point to inherent problems with the movement: it promised too much to too many; it easily disappointed converts; its emphasis on rapid technological change didn't appeal to farmers' more conservative approach to technology adoption.[3] Chemurgy was a longer-term program, and for farmers, the New Deal offered more immediate subsidies; the goal of self-sufficiency also lacked appeal for farmers who profited from selling globally. Moreover, agricultural industries aren't lucrative for investors, because they have downtime in between harvests and only have raw materials available once a year, making chemurgy's progress slow.[4] There was also active lobbying against chemurgy by the petroleum industry. And there were problems internal to the movement, such as personality conflicts at the top and a narrow dependence upon private funders. In the assessment of historian Randall Beeman, chemurgy was guided by corporate-technocratic scientists in the employ of industry, rather than by agrarian leaders.[5] Finally, its proposals for technocratic reapportionment of nature's resources and "sunlit lands" sometimes took on dark neocolonial and racist undertones.

What if chemurgy had emerged not in the embattled late 1930s, but today? In fact, is chemurgy back? "Chemurgy is returning with vengeance," asserts management professor Quentin Skrabec, "but don't look for the term to return. Today it is called biotechnology, ecology, green, or bioengineering."[6] Early chemurgy promoters like Ford and Dow Chemical are on the forefront of the return of chemurgy. But the first version of chemurgy, the eccentric path-not-taken, is worth keeping in mind as proposals for expansion of the bioeconomy, as well as the Green New Deal, are floated in response to climate crisis. The failure of chemurgy showcases the challenge of mounting a technologically rooted response to complex social problems. It illustrates how the dark side of the promoters and their ideology can muddle their efforts and legacy. It also highlights how elite commitment to fossil fuels has a history of thwarting alternatives, even when the alternatives come from within established industrial interests.

Technologically, the vision of a bioeconomy filled with renewable, plant-based goods should be easier than ever to attain. But socially, it seems even more distant. Right now, chemical firms are investing in small innovations, while chemical innovation on the scale of the postwar era seems too risky and long term.[7] Within chemistry and materials science, it could be that at the moment when we need bold, strategic thinking, we have a form of capitalism incapable of handling the temporalities at hand. Yet there are interesting scientific and cultural currents at work within biotechnology and agriculture. In Part I of this book, I'll discuss how some practitioners of cultivation are transcending conventional agriculture toward a kind of carbon alchemy: a sophisticated transmutation of life and land for the ends of sequestering carbon, in laboratories as well as in fields and seas. Like the philosophers of yore, they are innovating new recipes and tools to change matter from one thing into another. Their aim is to take the carbon that is heating us up and put it back into a regenerated earth system.

We'll begin by looking at a technique that has been included in integrated assessment models of how the world could decarbonize: bioenergy with carbon capture and sequestration (BECCS). In order to reach two-degree-Celsius targets, these models assume significant amounts of BECCS. The idea of BECCS is that a chain of actors grow biomass, burn it in a power plant that can separate out the carbon, and then transport the carbon somewhere to be stored underground. This carbon-storage part is key. Without it, the system is just regular old biofuels, which don't remove net carbon from the atmosphere. Carbon capture and storage is a well-established technology (and one that we'll discuss more in Part III of this book). Because both bioenergy and carbon capture and storage are known, BECCS sounds doable: doable enough technically that it was factored into the models. But it may actually be the least likely of these carbon removal techniques to be implemented.

Addis Ababa, Ethiopia, May 2013, 25°C / 77°F

The road teemed with blue Lada taxis in the hot, dusty night. Exoskeletons of rickety wood scaffolding clung to the rising skyscrapers. The China Road and Bridge Corporation had just opened the gleaming black Bole Road with a plaque celebrating China's friendship, and it felt like the place to be. This was just after the biofuel boom.

I was in a bar, drinking Ethiopia's most popular beer, St. George, with a couple of foreign correspondents. They had come off of a long day's work helicoptering to the far western reaches of the country, courtesy of the Ethiopian government. The government had orchestrated a ceremony for the divergence of the Blue Nile, as the Grand Ethiopian Renaissance Dam had reached a milestone in its construction. The Blue Nile is one of the Nile's tributaries, snaking its way down from the Ethiopian Highlands to Sudan, where it meets the White Nile. The dam will be the largest hydroelectric facility in Africa, generating a projected 6,450 megawatts and carrying a price tag of around $5 billion. That's a lot of power, considering Ethiopia's current electricity generation is around 4,000 megawatts, for one hundred million people.[8] But there are ongoing fears in Egypt, and from analysts around the world, about how it will impact Egypt's water supply, given that 90 percent of Egyptians rely on the Nile.

Ethiopia's appetite for grand-scale dams intersected my area of research interest—large-scale land grabs for biofuels—and so I was keen to talk to these correspondents. In particular, I was seeking information on how the rush for land for feedstock was impacting water and people. The water from Ethiopia's other new mega-dams, the Gibe series on the Omo River in the country's south, was advertised to investors as enabling large-scale irrigation on land they could lease. Foreign companies were being recruited to till Ethiopian soil and export the wheat, flowers, and oils.[9] This would help the government receive much-needed foreign currency. If you read about these deals in the press, you'd think they were being handed out like candy. Some deals were inked for cultivation of food crops, others for biofuels. Reports by NGOs like the Oakland Institute or tracking sites like the Land Matrix put the newly leased acreage in Ethiopia at over 3 million hectares, an area close to the size of Belgium.[10] Analysts responded with alarm as a cycle of information flowed between NGOs, the press, and academics, who saw their worst fears of accumulation by dispossession playing out. "The current land acquisitions in Africa can indeed be termed a new 'scramble,' because influential and wealthy foreign powers hasten to acquire land in order to secure their interests (future fuel needs and food market demand abroad)," wrote one scholar.[11] But what was happening on the ground? One could track down heavily photocopied leases to companies with names like Agropeace Bio or Saudi Star, stamped with the Ethiopian star and words in Amharic. But where were the products of cultivation, and where were they flowing? Did any of these biofuel plantations actually exist?

From afar, I had been tracking down scraps of evidence from the Internet: a YouTube clip of Ethiopian men and foreign investors touring flat fields with spiky castor plants; a chance press report from a journalist who was actually standing in a field, instead of just a press release announcing a new deal or memorandum of understanding. The reports from the province of Gambella, a lowland, infrastructure-poor region on the border with Sudan, were that 35,000 households had been "villagized" to make way for land deals—meaning people were forced off their land and placed into settlements, often in places that were forested or uncultivable. Socially, something terrible seemed to be happening related to these land acquisitions. But in terms of seeds in the soil, it was unclear what exactly was going on.

Understanding how the global land rush intersected the last decade's biofuel boom can help us see why projections around the scale-up of bioenergy with carbon capture and sequestration are so concerning. In 2007 and 2008, food and oil prices both spiked. Assessments vary as to the causes for the underlying price spikes: financialization of food markets and weather played some role, as did diversion of food crops (soy, maize, sugar) and land area toward biofuel production. Moreover, after the 2008 financial crisis, investors were seeking to diversify into a more real asset class. The "population bomb" narrative, popularized by biologist Paul Ehrlich in the 1960s, also reared its head again, never having really gone away. Hedge funds, seeing food prices projected to be high for decades, and relatively low farmland prices in many parts of Africa, believed the land to have a great potential for capital appreciation. They viewed land as offering stable returns and as a hedge against inflation.[12] With soaring food and commodity prices came an appetite for land from even more kinds of investors: sovereign and pension funds, as well as conglomerates from sectors like energy, agribusiness, and chemicals.[13]

Though the rush for fertile land was a worldwide phenomenon, two-thirds of the demand was in Africa.[14] Much of the land in Africa is owned communally, or in the case of Ethiopia, by the government, and documentation of land ownership can be rare. Ethiopia in particular was seen as a good place to invest: the developmentalist Ethiopian state would facilitate land deals for agriculture, and the dams were its proof that Ethiopia was part of a powerful future, inscribed in hard infrastructure. In Ethiopia, a great many of the proposed deals related to cane, castor, jatropha, and other biofuel crops. The acres of green were to be a clean source of prosperity both for firms abroad and for Ethiopian farmers; a win-win. The tenuous vision of this part-hopeful, part-apocalyptic future was what brought me to Addis.

But in most cases, there was no evidence that these deals had actually resulted in any cultivation whatsoever. The journalists I drank with that night in Addis had experience on the ground investigating one of the most notorious deals: the lease of 300,000 hectares by an Indian firm called Karuturi. Karuturi had hoped to grow and process corn, sugar, and palm oil. This was the largest deal, and the one that glossy NGO reports profiled in their sidebars; the poster child for a land grab. In fact, the lease was renegotiated down to 100,000 hectares, and just 5 percent of that new lease had been developed when we spoke back in 2013. There were all kinds of problems, including the fact that the leased land was on a floodplain.

Given the failures of so many of these schemes, I asked one of the weary correspondents: At what point does the government say this isn't working, we're not doing this anymore? He replied that the government would never give up, because these are long-term projects, and they have a long-term vision. Certainly, the government was interested in getting foreign cash; the large-scale land leases were a way to do this. But the Ethiopian government is also quite canny, and they weren't just giving out land wildly. The leases had stipulations about actually putting the land into production and not just speculating on it.

While in Ethiopia, I talked with several experts—an agricultural researcher here, a young former McKinsey consultant there—who told me of the difficulties of actually setting up a large-scale enterprise in a landlocked country with poor infrastructure. Outsiders tended to blame the people, government, and the land itself for the failures of these ventures. *They just don't want to weed. They did all the wrong spacing. The soil is deficient in six micronutrients.* It also turned out that crops like castor, a spiky oilseed plant that grows like a weed and is famous for requiring little water, are only truly profitable if you irrigate them. The main thing frustrating the ambition of biofuel production, though, was the reprise of an old story: oil prices went back up. In short, in a landlocked country without the necessary infrastructure, few or none these companies had managed to produce abundant commodities. A few years after our conversation, in 2015, Karuturi's lease was canceled. And in a 2017 letter, the company stated: "We stand tired and defeated and wish to exit Ethiopia."[15]

Failed deals don't mean that "nothing happened," though; speculative "phantom commodities," as social scientist Benjamin Neimark and colleagues have dubbed them, can shape the landscape and social fabric for years to come, even though they were never planted.[16] In developed countries, ethanol has continued strong, but companies who had reaped millions through investment in advanced biofuels, like algae, started to pivot to manufacturing other products. One designer of algal bioreactors for fuels is now in the business of "heirloom cannabis varietals."[17]

Biofuels were not an utter bust worldwide: indeed, biofuel production and use targets were met in some countries, reducing import dependence and substituting for demand for fossil fuels in Brazil, the United States, and Thailand.[18] Plantation jobs for labor-intensive crops appeared in some sites, generating income for some smallholders. But in general, biofuels have not met expectations around creation of long-term, high-quality jobs, alleviation of poverty for the most disadvantaged farmers, or improvement to energy access in remote rural areas.[19] In the commercial palm oil plantations of Indonesia, anthropologist Tania Li writes, there emerged a routinely violent and predatory system for capturing plantation wealth, in which regulations, rather than protect, served as points to extract further tolls.[20] In short, for biofuels thus far, the benefits have not been spread around, and the harms run rampant over many forms of life.

The scarcity of land—real or even just perceived—can transform lives and livelihoods. This is the first reason why BECCS systems for carbon drawdown seem unfeasible—land for feedstock cultivation could require 500 million hectares, an area one and a half times the size of India.[21] Dedicated bioenergy crops also require substantial use of inputs like water and fertilizers, implying both conflict over resources and increased water pollution from fertilizer runoff. And there are still more challenges with BECCS systems. Bioenergy with CCS would be competing with natural gas with CCS, since the CCS technology and costs are the same for these two fuels.[22] Moreover, with BECCS, you need to grow biomass near a place that the carbon can be stored. This is more than a minor logistical point: Biomass feedstocks are bulky, with relatively low energy density, and it's expensive and inefficient to transport them all over the place. And geological sequestration opportunities are not found everywhere. Thus, any such scenario assumes a massive, sprawling transportation infrastructure for biomass and for carbon dioxide. And finally, there need to be experts to cultivate the biomass—farmers who are motivated to grow those feedstocks over other crops. Farming is an art and science, and expertise and interest on the part of farmers is crucial to commercial viability.

The design choices in a BECCS system are everything. It is fiendishly difficult to grow biofuels in a carbon-neutral way when commodity chains are designed for low costs rather than low carbon. The concept behind bioenergy is that plants absorb carbon as they grow, which is then released when they are burned: So isn't that carbon neutral? Alas, biomass as a fuel source is not inherently carbon neutral, and can even be worse than fossil fuels. To calculate whether biofuels are carbon neutral, you have to account for (1) how much carbon was lost by cutting down whatever was there before; and (2) indirect land use changes (i.e., if a farmer decided to grow biofuels, and then a forest elsewhere got converted to food production). In general, it takes biofuels a very long time to pay back this initial cost, to make up for the loss of carbon that was in the landscape before the biofuels were planted.[23] Here's an example of poor system design: trees from North Carolina and other southern US states are shipped to the UK and burned, because wood has been designated a carbon-neutral fuel by EU regulators. In theory, the trees grow again. But some research has shown that wood-burning plants can have higher net carbon emissions than comparable coal plants for their first four or five decades of operation.[24] Investing in one energy source, like wood pellets from another continent, risks missing out on opportunities to invest in something better. To be clear, though, BECCS would likely have a better life cycle analysis than a typical biofuel plant, because the former captures the carbon dioxide emissions for storage elsewhere—tipping the analysis toward carbon negative.

Taken together, these issues make management of the earth's carbon cycles through crop cultivation seem like a dismal prospect—at least under the current unsustainable system of industrial-capitalist agriculture. Moreover, to some tech analysts, the failure of advanced biofuels to develop at scale and the prospects of the vehicle fleet electrifying make biofuels seem like a dated idea. So why does this concept still have any life in it? Possible answers include (1) that because it was the powerful farm lobby that posited them, biofuels have become a lumbering machine, an institution that simply stumbles on; (2) that modelers needed a fix for the models, and BECCS seemed the most plausible; and (3) that "drop-in" fuels—those that can be switched out for fossil fuels—remain something of a holy grail, and biofuels are the most familiar of these (as opposed to newer synthetic alternatives).

These twin canvases are the backdrop against which BECCS emerges: first, one of hope for the bioeconomy, to which BECCS could give certain life; and second, a backdrop of broken remnants, of inefficient and harmful biofuel experiments in places like Ethiopia and beyond. It is a strange, dissonant place to work from. Given this dim picture, to imagine BECCS seems, on one hand, an exercise in self-delusion. Indeed, many civil society groups have tagged BECCS with labels like "myth" or "fantasy." On the other hand, is it possible that this first generation of biofuels could be followed by second, third, and fourth generations of a different nature? Fostered by different forms of social organization, could such "Cinderella" biofuels bloom, marrying CCS in some kind of union that enables a stable climate?

Rebooting biofuels

Cyanobacteria were the first life-forms to convert sunlight into energy. They were also the perpetrators of the most significant mass extinction in history so far: the oxygen crisis 2.4 billion years ago. These little creatures are truly standout organisms. Called "blue-green algae," they're not actually plants—they're bacteria that photosynthesize, making their own energy and exchanging CO_2 for atmospheric oxygen. They manufactured about 30 percent of the oxygen you inhale.[25] They also have their own circadian clock, meaning they experience their own version of jet lag if you transport them across time zones. The wonder doesn't stop there. Scientists are engineering them to fix more carbon, thus making them a more efficient source of biofuel. And cyanobacteria are just one of many varieties of new-wave biofuels that could potentially be used for a BECCS with fewer negative impacts.

New technologies could, in fact, improve the picture for BECCS (though I will argue that we need both new technologies *and* new forms of organized production to make BECCS actually possible). You may have heard of "waves" or "generations" of biofuels (first, second, third, fourth), which refer to innovations both in feedstock and in processing technologies. First-generation feedstocks, such as sugarcane and vegetable oils, competed with food crops. BECCS in model simulations is assumed to rely on second-generation energy crops, which include nonfood plants. For example, grasses like *Miscanthus*, trees like willow or poplar, and crop residues like wheat straw or woody biomass are all second-generation feedstocks. These advanced fuel crops often contain more energy, and they can grow on "marginal" land. Some of them are "cellulosic" biofuels, meaning combustibles can be produced from cellulose, the fiber of the plant.

The world has been waiting for these cellulosic biofuels for years. There are a handful of cellulosic biorefineries in development, and they've received tax credits and infrastructure grants. Official targets have reflected high expectations for cellulosic biofuels in the United States: when a 2007 law mandating them was passed, it established a target of 11 billion liters per year. The Environmental Protection Agency has been steadily revising these targets downward. How much cellulosic ethanol is the United States actually producing? In 2015, it managed just 8.5 *million* liters—a far cry from 11 billion.[26]

One senior scientist I spoke with explained that research into cellulosic ethanol started twenty or thirty years ago, but it only became clear in the late 2000s that cellulosic ethanol was going nowhere, even after a billion dollars was poured into research in just five years. Why have cellulosic biofuels been so slow to materialize, despite this billion-dollar effort? One easy answer, again: cheap fossil fuels, braced in part by the rise in hydraulic fracturing, or "fracking." Another point is that it's simply very challenging to engineer these fuels. Plants are designed to stand up; thus, their cell walls have a tough polymer called lignin. To make fuel out of a woody plant, you need to be able to break down the lignin. You can do this using a thermochemical process (extreme temperatures, high pressures) or biochemically, with enzymes. But these enzymes that are quite expensive. It is therefore very hard to produce these fuels cheaply.

What about third-generation fuels? In some instances, "third generation" includes any biofuel designed or tailored for higher efficiency, such as low-lignin trees. In other instances, "third generation" refers specifically to algae, which gets its own category because of its higher yields, and its versatility. Algae have rich genetic diversity: there are between 1 and 10 million algal species to explore. They're also twice as efficient at using sunlight as many other crops, and up to half of their biomass is lipids (i.e., they are high in oil). When algae get contaminated, a crop failure only lasts a matter of days. Industry enthusiasts portray algae as "sunlight-driven cell factories" that convert carbon dioxide to potential products such as pigments, fine chemicals, bioactive molecules, biofuels, and more. One way to think about algae biofuel research is as domestication of wild species. Traditional selection took thousands of years, but now we can do this in years, going from "fit for purpose" to "design for purpose" crops—part of a "brand-new agriculture" offering food, fuel, nutraceuticals, and more.[27] And the terrain has barely been touched—right now, only about fifteen of those millions of known microalgal species are cultivated in some form, and only a few of these "domesticated" algae are cultivated at large scales.[28] So this is often posited by entrepreneurs as a new project (despite it also being a 1970s project). Algae cultivators find it exciting to basically get to do agriculture again from scratch, and by design. Algae could also be used with second-generation biofuel feedstock carbon capture in an integrated BECCS system. For example, one study proposed replacing soybean fields with algae and eucalyptus forests planted together: the biomass from the eucalyptus would provide the algae with heat, carbon dioxide, and electricity, with the remainder of the carbon stored. The system as a whole would produce more protein than soy does, with no increase in water demand.[29]

With fourth-generation biofuels, the generational terminology starts to break down. To some, "fourth generation" refers explicitly to the goals of carbon capture—either in the engineering of the feedstock, or that of the processing technology. One approach is to engineer biofuel feedstocks to make them more efficient at capturing sunlight, and eventually at producing fuel. This generally involves (1) improving photosynthesis, either by increasing light-gathering capacity or by extending the range of the light spectrum the organism can utilize; (2) improving carbon fixation; or (3) increasing oil content. For example, a collaboration between Synthetic Genomics and ExxonMobil made headlines in 2017 for using the gene editing tool CRISPR to modify an algae strain to enhance the oil content from 20 percent to above 40 percent, accomplished by fine-tuning a genetic switch that regulates the conversion of carbon dioxide to oil in *Nannochloropsis gaditana*.[30] To be clear, these aren't carbon removal techniques unless the carbon is sequestered and stored, but it's possible to imagine BECCS plants that use genetically modified algae as the feedstock.

Another conception of "fourth generation" fuels incorporates synthetic biology—the design and construction of new biological entities. One approach is to engineer microbes to produce biofuels without feedstock—going straight from microbes to fuel. For example, photosynthetic microbes can be engineered to excrete biofuel or biofuel precursors. Startup accelerator Y Combinator recently issued a request for proposals from carbon capture startups working on enzyme systems that don't involve organisms, known as "cell-free" systems, which would be engineered to synthesize carbon dioxide into other compounds without the construction of a new species.[31] Cleantech company LanzaTech aims to turn "our global carbon crisis into a feedstock opportunity" by engineering a bacteria called clostridia to make fuels out of carbon dioxide. Clostridia, which is found in rabbit droppings, can fix carbon from carbon dioxide and feed on it. (Synthetic biology also has applications for non-biofuel-based carbon removal; for instance, scientists have just begun looking at how engineered bacteria could be placed into CCS sites to speed up the conversion of captured carbon into calcium carbonate.)[32] Synthetic biology approaches to carbon removal that are effective and rapidly scalable may have the potential to circumvent the aforementioned constraints of BECCS. Perhaps these technologies would even become successful enough to obviate consideration of solar geoengineering.

Through exploring these ideas of *better* and *more efficient*, we have traveled several levels of abstraction away from the bioeconomy. Biological approaches for problems like climate change and energy are not so easy or elegant after all, given that they involve reining in the messiness of life. Interestingly, there isn't a popular umbrella term, movement, or unified field around engineering a better photosynthesis or a biosynthetic fuel—it just seems to be something increasing numbers of scientists in both academia and the private sector are interested in, even after the algae boom-and-bust. The key thing, though, is that much of this engineering is being done simply to get around these constraints of profitability, to become cost competitive with fossil fuels. What if those constraints were changed?

What would biofuels look like without capitalism?

It is clear that new biofuel technologies—whether they be cellulosics, algae, or solar fuels—are not going to "naturally" replace first-generation fuels. Substantial social and political pressure would be necessary to develop biofuels for carbon removal, and their deployment at climate-significant scales would be a massive feat of social engineering. This is implied in a reading of the scientific literature. For example, environmental scientists Mathilde Fajardy and Niall MacDowell state that the sustainability of BECCS relies on intelligent management of the supply chain. In a sober, not-at-all politically inflected article, they identify five key levers to make BECCS actually carbon neutral or negative: "(1) measuring and limiting the impacts of direct and indirect land use change, (2) using carbon neutral power and organic fertilizer, (3) minimising biomass transport, and prioritising sea over road transport, (4) maximizing the use of carbon negative fuels, and (5) exploiting alternative biomass processing options, e.g., natural drying or torrefaction."[33] These five levers each imply a truly progressive politics, and without movement on these levers, BECCS would never be climatically significant. But it is possible to imagine moving the levers by bringing a different politics into the picture. If production were in the hands of the people who live on and work the land (managed, for instance, via agricultural collectives rather than exploitative contract-farming arrangements), and if the people owned the resources for production (the seeds as well as the cultivation and processing technology), they could choose to use them on lands that they know are *actually* neglected or marginal, as well as intercrop these lands with food crops. There's plenty more to say on this, but one point is key: to be truly carbon negative, BECCS would require a totally different social logic.

Sketch: Flowers

I gather flowers as I walk. A sprig of lavender, a scarlet trumpet, scraps of our landscape for her.

It's hot. A dry heat. The only sound is the quiet hum of the pumps under the algae raceways, slowly stirring the bubbly viridian mats.

There's a canopy of grapevines, with a rustic bench that I've never bothered to sit on before. The gnarled slab of bristlecone pine, under the parabolic curve of the canopy, has my mother's touch. She likes volumetric, ambitious shapes, but to garner votes for her landscape designs, she tries to balance them with these organic flourishes. Knowing her, this stubbly bench feels like a concession.

My mother worked on this part of the village, from the plaza up to the solar orchard, with its scruffy sheep roaming around under the panels. I had lobbied for goats as a kid, but she overruled even bringing that idea to committee. Goats are trouble.

The lab is visible on the ridge; a low building lying in the shadow of the biorefinery. Chira works there. She's not from here—we met on a collaboration platform. We were interested in the same microorganism, and we started trading notes, making up in-jokes about it. It was luck that a job opened up at a lab in my town right when she was graduating.

It's true: I'm dawdling on this bench. She might never want to see me again.

I rearrange my droopy flowers and follow the stream. It's lined with shrub willow, all the way down to the treatment plant. We used to have to make baskets with them in school. Mine always had sticks jamming out from every angle. My brother, on the other hand, made baskets that are still storing herbs around my parents' kitchen. Now, he lives from his monthly ubi and makes pots. I have collected a cabinet of his pots, and whenever someone at work has a birthday, I've got a pot with their name on it.

Little lizards dart across the path. I remember how entranced Chira was with this path, and the tiny yellow LEDs lining it. "The plants need water," I had explained to her, and paused our walk to step down to the stream, scooping up cold water and drenching them. Soft chime, and the lights went green: she clapped her hands. "There used to be a game where you'd get points for watering them, and then the council would give you a T-shirt or something, but then people decided that was dumb and they'd just water them anyway." I was trying to show off what a good citizen I was. She was more impressed with the path, and the green-and-pink hummingbirds swooping around. But we had had our first kiss on that walk.

Here's what happened. Chira and I were having dinner at my place. I had made a garden vegetable lasagna, the candle was dripping wax all over the table, and it was warm.

We were listening to the crickets, savoring the evening, I thought, when she said, "How would you feel about going to Kobane?"

"Kobane? There's nothing there but dust." I mean, I had read the Wikipedia article when we started dating, but I had never thought about going there.

She folded her napkin, placing it neatly on the table.

"Wait, why? Are you going back to Kobane?" I asked.

"Well, it's part of my apprenticeship agreement that I would move back to the region, and contribute what I've worked on." Chira sat back in her chair, and crossed her arms.

"I know that's technically the contract, but I don't know anyone who actually follows through with those."

"You hardly know anyone besides people who grew up here. You've never even been beyond the borders of this region."

"That's not true," I protested. "Anyway, what would you do there?"

"They're setting up a co-op biorefinery. They need people to train students who are optimizing for the specific plants in their fuelshed. Some of these native crops are still pretty under-studied."

"Sure, but anyone could do that."

"It doesn't work as well if some outsider comes in and does it. It needs to be someone that speaks their language. Anyway, it's clear that you don't see anything of value there." She got up and grabbed her bag.

I stood up. "Wait! I'm sure there's lots to do in the area. I mean, there's desert wildlife, mountain climbing … There's mountains, right?"

The door clanged shut behind her.

I started cleaning up the dishes, washing lasagna debris down the drain, but I couldn't stand the screeching of the crickets in the night. I walked across the village to my brother's place.

Jorge lives with eight others in an ubi-house in the center of town. He earns enough from his custom-designed pots that he has his own room, instead of just a bunk. This is good, because I happen to know he snores. I walk in without knocking.

He was lying on his bed, reading. "What's new?"

"I think Chira just dumped me." I recounted our dinner to him.

He winced when I get to the part about the dust. "You're a dick."

"You have to admit it's true, though. They totally mined their aquifers and turned a grassland into a wasteland. It's horrible."

"Have you ever been there?"

"No."

"Well, maybe checking it out in person is step one."

"You've never been there, either," I replied. "You barely even leave your house except to go to the studio."

"Okay, but listen. You're too young to remember grandpa always going on about respecting the travelers who come and live with us."

"But I do respect Chira."

"Sure, you respect her. You respect her journey. But you don't respect where she *comes from*. As an actual place, with actual people," Jorge added. I picked up a one-legged clay heron that's sitting on his desk. Jorge sat up. "Don't touch that."

"I guess it's true that I don't respect where she's from. I mean, I don't know anything about it—how could I respect it?"

"Why don't you know anything about it? You've had a year of Chira's company to learn something. Didn't you ask?"

"No."

Jorge sighed. "You think you're so intellectually curious because you went out and got trained for a job about plant genetics. But if you'd spent some time lying around here reading"—he gestured to his tablet on the nightstand—"then you might know something about what's going on over there. It's not like it got dusty all on its own. The world stood by and burned more carbon while it dried out, and they turned a blind eye and let the oppressors just mow everyone down. It's amazing they are even growing anything out there. Really a miracle."

"I didn't come here for a lecture." The heron was staring at me from behind its pointy beak. "What am I going to do? I wish I could just take back everything I said."

"The problem isn't the words you said. It's what it revealed about your character. You can't take back that revelation. She sees you in a new light, now. All you can do is damage control, by acknowledging your misdeed and explaining what you're doing to become a better person."

I sighed. "Maybe I *should* go to Kobane."

"You could, but if you do it just for her, you're going to be miserable. I'd be surprised if she wants you along at this point." Jorge started throwing a ball against the wall. His attention had already wandered from my plight. "Hey, do you want a smoke?" he said, taking out his rolling papers. He knows I don't smoke.

I stormed back to my house and spent all night reading and watching virtual reality vids about life in her region. Then I thought about how *good* she is. What her childhood might have been like. I know the expression on her face when she gets a good idea, and I know what will make her laugh. But maybe that's all I know.

The trail up to the top of the ridge is rugged. Hardly anyone comes along this footpath. From up here, I can see the village, the plaza, the green fields and shiny solar orchards and forests beyond—almost to the edge of the fuelshed. The biorefinery is just below the ridge, the railroad tracks and roads sprawling out away from it. The numbers on the biorefinery's wall have just changed. It's made of minimalist LEDs, most of them glowing the color of wood (even though the facility is made from high-carbon concrete—tacky, if you ask me). The eraser reads: *56,201,008 tons removed.* I'd never really even paid attention to it, but Chira had noticed it immediately. I'd explained that the idea is to decommission the carbon capture part at half a billion tons. At that point, they'll take apart the pipelines that transport the carbon dioxide away. "Not in our lifetimes, though," I'd said, wrapping my arm around her.

So here I am in the bike lot, with my spindly flowers, among the bicycles and workers heading home. I run my fingers through my hair. Large puffy clouds are drifting overhead, marching out across the hills. Chira's coming out of the building, now, the glass door revolving. She sees me, and pauses briefly. I can see her roll her eyes.

2

––––––––––––––––––––––––

Cultivating the Seas

Within the next fifty years fish farming may change us from hunters and gatherers on the seas into "marine pastoralists"—just as a similar innovation some 10,000 years ago changed our ancestors from hunters and gatherers on the land into agriculturists and pastoralists.

—Peter Drucker

I f earth's lands are full and used, where else can be cultivated? While some dreamers in Silicon Valley and beyond hold out hope for eventually moving offworld, others are looking toward the oceans. To enthusiasts of ocean colonization, we are in an epochal moment: one in which domestication claims a new domain. "We've pushed agriculture and the Green Revolution to its limits on land, but remained hunter-gatherers on the ocean," writes the Seasteading Institute, declaring, "A Blue Revolution in ocean farming technology would launch seasteads to center stage."[1] But you don't have to be a libertarian who dreams of living offshore on an independent floating "seastead" home to be excited about the prospect of a counterpart to the Green Revolution of the mid twentieth century—or even a new form of marine civilization. An apocryphal Jacques Cousteau quote floats around the Internet: "We must plant the sea and herd its animals … using the sea as farmers instead of hunters. That is what civilization is all about—farming replacing hunting." Technological breakthroughs in cultivation can enable this shift to farming in the blink of an eye, geologically speaking. It took 6,000 years of cultivation to transform the wild grass teosinte into modern-day corn, but domestication of marine species such as seaweeds and marine microalgae can happen within decades.

Marine cultivation approaches to carbon capture are alluring because the deep ocean is one of the remotest places on earth, and when carbon reaches it, the carbon theoretically could stay there undisturbed for a very long time. A recent study in *Nature Geoscience* estimated that seaweed floating down to the depths of the ocean naturally sequesters 173 million tons of carbon a year—about as much as New York City emits.[2] To increase that sequestration, though, macroalgae would have to be turned into bioenergy, with its carbon separated out and stored, in a BECCS system that uses macroalgae as its feedstock—or some other more inventive way to store the biomass reliably in the ocean depths would have to be devised. But some scientists find the idea worth studying, seeing in the extreme productivity of marine environments an opportunity for carbon capture that progresses more rapidly than on land: kelp can grow two feet in a day. To learn more about what it takes to cultivate seaweed, and what opportunities this presents, I paid a visit to a seaweed cultivation lab.

More than a new superfood

I step into the chilled room, irrationally reluctant to let the heavy insulated door close behind me in case I somehow get trapped in here. On the left side of this cavernous refrigerator / plant growth incubator are shelves bearing glass jars filled with water; on the right are flasks, lit by fluorescent tube lighting, temperature-controlled. All of the jars and flasks have tubing threaded into them, and are quietly bubbling away, like futuristic orbs, each bearing a strip of blue masking tape scrawled with Latin names. I kneel down to peer at the material floating inside. Some jars hold delicate brown flakes, curling around the edges, mushroom-like. Others have angry little spheres, or dark crimson tangles. Each flask seems to be inhabited by a different personality.

This lab, at a campus of the University of Connecticut near Long Island Sound, is where marine biologist Dr. Charles Yarish and his lab group study seaweed. Yarish has spent his forty-plus year career nurturing the emerging seaweed industry in the United States. While seaweed still remains an overlooked crop in the United States, elsewhere in the world, production has skyrocketed. Red, green, or brown seaweed; *Eucheuma*, *Porphyra*, or *Gracilaria*. Domestication of macroalgae began only in the twentieth century, but already one hundred species of macroalgae are produced for a market now worth over $6.7 billion.[3] There's been a sixfold increase in production over the past twenty-five years, mostly in China, which is responsible for over half of production, and in Indonesia, responsible for another third. (US seaweed production doesn't even make it onto the charts.) The stuff is prolific and is able to attach to hard structures, meaning it can be cultivated wherever you build one. In Europe, wild seaweed is harvested, but in Asia, the main source is cultivation. In places like China, Korea, or Japan, rope lines with seedlings of hatchery-grown brown algae are suspended from floats in the autumn, grown through the winter, and harvested throughout the spring.[4] One can see these impressive installations from space.

In Yarish's office, beneath cabinets boasting bumper stickers like "I brake for algae" and newspaper clippings celebrating his seaweed cultivation triumphs, I show him something I've found in the archives. It's a little book entitled *Alchemy for the '80s: Riches from Our Coastal Resources*, which details a vision of scaling up a seaweed industry in the wake of concerns about fossil fuel resources.

"That's an oldie but a goodie, right there," Yarish says. "We're in the same place, and going back to the same issues."

In the 1970s and '80s, soaring fuel prices provoked a brief but strong interest in the possibility of offshore cultivation of biofuel feedstock. At the time I spoke to Yarish, he was more interested in food production—or other high-value products. "The lowest-value product you can get out of marine biomass is biofuels. Absolutely the lowest. But when you take a look at other applications ... if it's edible, it is a high-value product. Then there are other uses for seaweeds as well." He ticks off the phycocolloids, like alginate or carrageenan, which are used in antiaging and other cosmetic products; nutraceuticals, stuff you'd find in the vitamin aisle; biomedical applications, like antitumor and anticancer compounds. "Seaweeds, since they can't move, have to develop chemical defenses," he explains, making them a rich source of new discoveries of medically interesting compounds. The recent increase in production is driven largely by food demand, their main end use, though there are others: for example, the hydrocolloids that act as the skeleton of seaweeds can be extracted as clear, flavorless thickeners, which you'll find in ice cream, shampoos, toothpaste, baked goods, paper, fish feed, animal feed, and more.[5]

Currently, Yarish is taking a fresh look at developing techniques for mass production of seaweed with an eye toward biofuels. He received a government grant as part of the MARINER (Macroalgae Research Inspiring Novel Energy Resources) program of ARPA-E, the US government's Advanced Research Projects Agency for Energy (DARPA's progressive, green cousin). Projects funded by the grants include cultivation and harvesting systems, monitoring tools, and breeding and genetic tools—all technologies aimed at making the United States a global leader in seaweed production, aimed at eventually developing liquid transportation fuels. What separates MARINER from previous work in the 1970s–1980s is that it sees current opportunities in foods and feeds as steps that can lead to bringing down the cost of biomass production for biofuels.

From a biofuel standpoint, seaweed is an attractive feedstock because it doesn't contain lignin, the structure-giving component of plants that is expensive to break down. Seaweed as a biofuel would, of course, be only carbon neutral at best, since the carbon would return to the air when the biofuels are burned. However, if you also processed the feedstock in a way that would sequester carbon, you could have a kind of seaweed BECCS system. (Another way seaweed might mitigate climate change isn't as biofuels, but as cattle feed: one lab-based study found that adding a specific seaweed species to the diet of cattle—*Asparagopsis taxiformis*, which contains a compound that inhibits methane formation—could reduce cows' methane production by 99 percent. Scaling this cattle superfood up toward 2 percent of a cows' diet, globally, would take a sizable seaweed cultivation industry.)[6]

Because coastal areas are often biodiversity hot spots in need of protection—not to mention desirable for other human uses—attention has flowed toward growing seaweed biofuels farther off the coasts, utilizing the exclusive economic zones (EEZs) of nation-states, which stretch out to 200 miles offshore. However, as it turns out, it can actually be quite difficult to farm the open ocean. One example from history illustrates the challenges—and showcases the innovators still experimenting with this approach.

Ocean Food and Energy Farms

Capture sunlight, turn it into fuel: it sounds like a futuristic formula. Yet one of the pioneering efforts in kelp cultivation, the Ocean Food and Energy Farm project, began in 1972 with an infusion of US Navy funding. OFEF also gleaned funding from the United States Energy Research and Development Administration (the forerunner of the Department of Energy), the National Science Foundation, and the American Gas Association, which hoped that methane pro-duction on a significant scale could contribute to the national gas supply. (Just imagine a constellation of players like this interested in seaweed fuels today.) "If successfully realized, this technology would enable the planet's oceans to become a huge new source of feeds, foods, fuels, and chemicals—fixed carbon and fixed nitrogen —for the benefit of humanity," wrote the project leader, Dr. Howard Wilcox, noting that the major question was economic feasi-bility. "But" he added, "the issue is more one of 'when' rather than 'whether' the concept will eventually pay off."[7]

Wilcox's concept was to grow the giant kelp *Macrocystis* in the surface waters of the open ocean. Because most of these surface waters are a desert, nutrient wise, he proposed to pump water up from the nutrient-rich waters that lie about 300 meters below. But building the structures on which kelp would grow, as well as the mechanisms for pumping water up, presented immense engineering challenges. What's more, storms and biological organisms plague human efforts to cultivate this fluid terrain. How could the kelp be held together in a farm unit without being lost to tides and currents? To solve these problems, Wilcox imagined free-floating farms that would use propulsion systems to keep them circulating around the ocean's massive eddy patterns.

The scale was also a problem: to make fuel out of kelp, you need a *lot* of it. Dr. Wilcox's calculations indicated that the farms would have to stretch 100,000 acres or more to be a cost-competitive source of energy. Moreover, at the end of the 1970s, the gas sector was deregulated, and oil embargoes ended. And by 1982, the funders were no longer interested.

However, there has been remarkable progress in engineering in marine environments, and technology more broadly, since the 1970s. Could advances in robotics help kelp farms scale?

The startup Marine BioEnergy could be seen in some sense as a descendant of OFEF, as it was founded by the son of Dr. Howard Wilcox, Brian Wilcox, and his wife, Cindy. Marine BioEnergy is investigating cultivation of long-line kelp in robotic farms—work that has attracted press interest from the likes of *Fast Company*, National Public Radio, and the *New York Times*.

The new kelp farm concept is simple: the entire kelp farm gets moved down at night to receive nutrients from deeper waters, and up during the day to reach sunlight. This is a significant difference with the 1970s concept: submerging the kelp rather than building a large pumping infrastructure saves a lot of money because it means the farms can be smaller. Drone submarines would tow these kelp farms to new waters, communicating with harvesters by satellite, which would save labor costs. The drones could also submerge the farms to avoid big storms and passing ships. It's an ambitious but well thought-out project, involving many collaborators and supported by ARPA-E.

The first step for the Marine BioEnergy consortium is to figure out the mechanics of kelp farming. As Cindy Wilcox told me, "Right now the biology question is the dominant question: Does a fast-growing kelp thrive when submerged to depth to absorb nutrients and surfaced during the day to absorb sunlight?" Marine BioEnergy is working with marine biologists and scientific divers at the University of Southern California to better understand this piece of the puzzle. The various species will be tested on an anchored research buoy—also known as "the kelp elevator"—near Catalina Island. The buoy has a boom that is surfaced during the day and submerged at night to find out which species thrive in this environment. The target kelp, *Macrocystis*, grows a foot a day with adequate nutrients. Moreover, some of these kelp plants have been found to grow up to three times faster than their neighbors, implying that there is much more to learn about why some plants are more productive (something other researchers like Dr. Yarish are also working on).

The second step is to harvest the kelp and make a biocrude out of it. Researchers at Pacific Northwest National Labs have devised a process to convert kelp into biocrude (a hydrothermal liquefaction and catalytic hydrothermal gasification process) that takes about an hour in a reactor and doesn't require fermentation; in theory, methane output from the process itself could be used as power source. Then, an open question is whether it makes more sense to bring the harvested kelp back to land to process it, or instead to process it on the open ocean and have tankers come and fill up out there. Marine BioEnergy's latest idea is to rendezvous the drones with the harvesters four times a year. The technology for all of this exists, but it needs to be integrated for this new application, Cindy Wilcox told me. "Right now we are working to determine the most cost-effective methods to implement the subsystems and get the end-to-end process underway."

The fuel could also complement renewables, supplying both liquid fuels and methane: as Cindy Wilcox suggests, on days without wind or sun to power the grid, carbon-neutral methane could fuel the turbines. "We don't expect batteries or pumped hydropower to provide adequate storage on the scale needed at a cost-competitive price. Kelp-based methane pumped through the current pipelines will meet the need and stabilize the grid," she explains.

From ocean farming to carbon removal

The idea of cultivating seaweed went under the radar between the 1970s and today, but it didn't disappear. During the 1990s, there were several workshops held on the marine biomass concept—and they were specifically focused on climate mitigation. More recently, an interdisciplinary team has proposed a method for using seaweed not only to mitigate climate change but also to achieve negative emissions. Described as "ocean afforestation," the concept, outlined by marine botanist Antoine de Ramon N'Yeurt and colleagues in a 2012 paper, entails several steps: growing and harvesting the seaweed, digesting it as biomethane, extracting the separated carbon dioxide and methane, recycling the nutrients, and permanently storing the carbon dioxide via some geophysical or geochemical technology. (One idea is to store the carbon as liquid with seawater, inside a geosynthetic membrane tube resting on the seafloor.) They calculate that afforesting 9 percent of the world's ocean surface could be sufficient to replace fossil fuel energy, remove 53 billion tons of carbon dioxide, and increase sustainable fish production.[8] Some of the people involved with these initial calculators are continuing their work within a network called the Ocean Foresters.

Ocean afforestation sounds grandiose—but that may be because we don't have institutions that work on this kind of ecosystem-scale research and development. Most innovation is undertaken by narrow specialists pursuing particular niche products, whereas this idea evolved from a whole host of people, including a former wastewater engineer and a marine botanist from Fiji. Indeed, the authors note, it would be like putting a man on the moon, though likely a much better return on investment; it also involves low-tech components that could scale quickly. The Ocean Foresters have evolved their concepts in the years after the publication of this paper, and their ecosystem-level proposal has captured the imagination of popular science writing—including in the essay collection *Drawdown* (2017), or Australian author Tim Flannery's *Atmosphere of Hope: Searching for Solutions to the Climate Crisis* (2015). Mark Capron of the Ocean Foresters points out that it's not a pure carbon removal technology;

rather, seaweed cultivation is best viewed as holistic mitigation and adaptation. Nevertheless, there has been little actual momentum to pursue these types of multi-step, holistic solutions, which, in order to form even demonstration projects, require coordination of multiple scientific and engineering fields as well as among ocean users.

Industrial or boutique production?

Carbon removal at climate-significant scales with seaweed seems to imply industrial production: vast monocultures that can reap economies of scale. But many seaweed enthusiasts are interested in a different model, a polycultural approach to marine agronomy called integrated multi-trophic aquaculture (IMTA). It can be thought of as a counterpart to some of the agroecological practices on land, in which by-products from one species are recycled to become inputs for another. There are plenty of traditional models. For instance, in carp polyculture in China, mulberry was cultivated using nutrient-rich pond sediments, the leaves were fed to silkworms, and waste from silk production and processing was returned to fishponds stocked with Chinese carp.[9] And today, in Sungo Bay in the Shandong Peninsula in Northern China, there's a system where abalone feed on kelp, and their waste is then used by sea cucumbers; in turn, the kelp assimilate the waste produced by both the abalone and sea cucumbers.[10]

A small but dedicated community of academics, entrepreneurs, and NGOs is advocating for this type of aquaculture. In the United States, from Connecticut to Maine to Alaska to California, cottage cultivators are experimenting with how to set up a seaweed industry compatible with sustainability and social justice principles. One oft-profiled organization, GreenWave, advocates for "3-D ocean farming," where kelp-and-shellfish farms make use of the whole water column. It isn't an easy task: How can you stand up a seaweed industry according to social justice and sustainability principles, when industrial production elsewhere is so much cheaper? People in Europe are thinking about this as well: for example, colocation of seaweed production and wind farms in the North Sea (though one study found this totally uneconomical, as the seaweed would face heavy competition from Chinese production).[11] Dr. Yarish explains, "We can't employ technologies that you see in Asia, where labor costs are very low; we have to adapt." This means using techniques that require not ten people but one or two. Part of his approach to helping the industry get on its feet is to make all his work open source. "I made a decision a number of years ago that in order to develop a seaweed cultivation industry in North America—the biggest problem had been people trying to always do things quietly. Secretly … And I felt that at least the primary research should be open source, so people would not have that as a limitation." Yarish's team has a free handbook for cultivating seaweed, as well as a great six-part introduction on YouTube. But another fundamental challenge is that in Western cultures, there is no culture of seaweed cultivation. So how does seaweed production start in a community? You want coastal communities to have a "fair return on their investment," Yarish explains, in terms of their labor, their energies, their nurturing of the farm system. If outsiders come in and start showing people how to grow seaweed, or begin investing the capital, it might not flow back to those communities. At the same time, to quickly scale up a seaweed industry that could support big goals like marine biomass for carbon removal, there would have to be some diffusion of knowledge or technology. In the absence of strong government intervention, the seascape favors big companies—or otherwise, requires a coordinated, collective effort by farmers working together, the beginnings of which we can see in places like Connecticut.

Seaweed cultivation can come with environmental co-benefits, such as reducing commercial fishing activity and allowing fish stocks to recover.[12] In China, seaweed aquaculture is already of such a scale "that it may be of regional biogeochemical significance" —for instance, it's playing a pivotal role in addressing the problem of coastal eutrophication, according to a paper in *Scientific Reports*. Scientists Xi Xiao and colleagues found that the seaweed industry is already removing 75,000 tons of nitrogen and 9,500 tons of phosphorus from coastal waters each year. Just 1.5 times more seaweed cultivation would be able to remove *all* the phosphorus flowing into the seas,[13] though it would be hard to take up all the nitrogen. Capturing these co-benefits will to some extent rely on good system design. For the time being, seaweed cultivation, because it is a new industry—for the West, anyway—is mostly unregulated.

If you've never thought about seaweed regulation—I hadn't, before writing this book—you might be mystified: Why would seaweed need regulation? But there are matters like the spread of invasive species, or diseases. For example, a bacterial disease called ice-ice infects a red seaweed called *Kappaphycus*, turning its branches into ghastly white icicles. During the past 10–15 years, the disease caused millions in crop losses in the Phillippines and Indonesia, before it spread to farms in Tanzania and Mozambique in the Indian ocean.[14] There's also regulation to help consumers. In general, people eating seaweed might want it to be traceable and clean, reputably sourced. Policy aimed at seaweed could also aid farmers by helping them cope both with low prices (most of the value is added in processing rather than in growing), and with the impact of climate change (since seaweed farming is vulnerable to increasing storm activity). State-subsidized aquaculture insurance, such as in Korea, could also support farmers. Another regulatory challenge will be to figure out how to incorporate seaweed farming as carbon removal into climate policy, at both national and international scales. For now, it's tough for seaweeds to qualify as carbon sinks under the UN Framework Convention on Climate Change. The definition has been set up for trees—in terms of carbon turnover time—but with seaweeds, the carbon they draw down is easily decomposed and released again, and the turnover time is less than ten years (versus several decades for trees), excluding for the detritus that goes into the depths.[15] Of course, there are a variety of ideas about how to sequester the biomass—sinking it into the deep sea, into submarine canyons, and so forth, which researchers have termed "seaweed carbon capture and sink," or "seaweed CCS."[16]

Despite all the challenges, the dream of multifunctional seascapes near coastal communities holds powerful promise. The Ocean

Foresters' vision includes construction of restorative coastal infrastructure, enhancement of biological infrastructure like mangroves and dunes, and use of tensile fabric to make flexible breakwaters. The automated offshore production would be stewarded by human attendants: the submarine and the aerial drone enable "boutique forest management," generating a decent living for the people taking care of the machines.[17] Other designs involve the use of offshore wind turbines to anchor seaweed cultivation. The production of sustainable biofuels would also help meet the energy needs of island and rural coastal communities. This all might sound futuristic, but the future is rapidly crashing down on us. In short, if we are to avoid repeating agriculture's grave mistakes of the past in fluid territory, now is the time to advocate for sustainable worker and community-oriented models of marine crop production.

From the perspective of researchers experimenting with seaweed for carbon removal, there are two looming risks—on top of the more basic risk of economic unfeasibility, namely, that the policy and public interest needed to support and nurture it fails to grow. The first risk is climate change. All over the world, for example, warming sea temperatures are decimating natural kelp forests. One scientific report graphically describes the "urchin barrens" that are settling in where kelp forest used to be—these warm-water species mow down everything in their path. Apparently, they are "almost immune to starvation," living for over five decades. And when they get hungry, they destroy more: their jaws and teeth actually enlarge when they are experiencing severe hunger. The report offers more grisly details: when stressed, they form fronts that march across the seafloor, hunting for food.[18] And the voracious urchins are just one instance among the many ways that climate change makes all kinds of agriculture trickier.

The second risk for people working on seaweed is that some other innovation comes along and makes the first technology obsolete. We often think of carbon removal, and decarbonization more generally, as a portfolio or series of wedges, each of which will be necessary and appropriate in a particular social context. In reality, however, some technologies come at the expense of others. It's quite possible that some of the geologic or chemical techniques for removing and storing carbon would be more attractive to current actor-interest groups than the natural and cultivation-based techniques described in this section. This is something the farmers and cultivators of the world may not be so thrilled to contemplate. Even so, seaweed biofuels potentially offer a transformative fuel source, and it's worth thinking about how the industry might look at the end of the century.

Sketch: Ghost Bar

A sign at the horizon read *Ghost Bar*. The blue neon bobbed on the dark water.

On-screen, a clump of glowing green showed seaweed waiting to be harvested, and two red lights signaled farms in need of attention. A barnacle in the gears, maybe. Have you tried turning it off and on again?

Someone else could have that gig. She didn't need to be top of the repair charts, just to make enough that she could save a little. She wanted to stop.

She'd been on the boat a week. She frowned at the mirror, rubbed citrus oil into her skin, tied back her hair as she drew close. The geopolymer base floated on fiberglass tubes. Nestled between the bar and the barkeeper's quarters was a tiny greenhouse and a chicken coop. She tied up among the boats docked at the platform edge.

A blast of ocean air followed her through the heavy door. There was no one behind the bar. She held her breath. And then the bartender stood from below the bar and saw her, and smiled. She smiled too.

"How are you doing, Vilma?" he asked.

"Another day in paradise," she said. She pulled up a stool.

One wall was all window. She saw a palm tree swaying in its bolted bucket. The sky was streaked with magenta and violet clouds.

"Usual?"

"Surprise me."

He went into the back. A black cat wound its way around her feet. The bartender returned with sliced tomato and a fried egg on hot flatbread. "Fresh," he said. "The egg."

"Awesome."

He poured her a generous glass of gin, twisting a banana leaf into an umbrella.

"This is gorgeous," she said, admiring the food. He poured himself a whiskey. They touched glasses.

She couldn't think what to say. After her last breakup, she'd wanted solitude, to work on her music. Hence the marine systems repair course, hence the boat: the only solitary place in the world was the open ocean. But there'd been months of it now, and she had six compositions, none of which she was quite content with.

He told her stories: about his customers, about the chaotic week when the real-time translator had broken down. She asked after the health of the chicken. The biocrude heater came on. Someone put nueva-bhangra on the virtual jukebox. The lights auto-adjusted to warm tones.

They told each other what they missed about land, and what they were happy to do without, as night deepened around the bar.

"Have you been tracking this storm?" she asked.

"Not really. Your guys with the drone farms have way better forecasts than what I get. I only get the internationals, I'm not in state waters."

"It's a straight-up hurricane now. And it's turned."

"We don't call them hurricanes out here."

"I mean …" She scrolled. "There's a fair bit of variability, but it's pretty brutal." She held up her tablet. He said nothing. "Fuck. I've got to sober up and get around this thing." Most of the customers, she realized abruptly, had already gone back to the water for the night. The lighting was now soothing twilight tones.

"You should take a quick nap," he said.

"No, I just need an hour to get straight. Then I can autopilot south. Sleep en route." He made her an espresso with orange peel and cinnamon. She studied the forecast as he stacked chairs. The floor began to roll. "You ought to move," she said.

"It's a huge pain. That's why I've just been floating here. People know where to find me."

"People can find you by tracking your updates. Your fans are following you, anyway, for your pithy observations about your chicken and your sunset photos, while they dream of the offshore life. Come on. I'll help you prep."

Together they secured the fittings at the bar. In the greenhouse he undid plumbing and covered up loose soil. He checked the fuel tanks and she the batteries. They scouted for loose objects on the platform, in the moonlight. She could see no boats now, save her own.

In his room, the books spilled out of bookshelves. The bed was unmade. He saw her look at it, and she saw him see.

"I'll get the cat. Can you help with the roof?" She ascended through the ceiling hatch to unscrew the solar panels. It was already the small hours, and the rain was drumming. She came down and checked her tablet again. "Really not looking good. Where you headed?"

He booted up a terminal to plot a course. "I'm not going to avoid it all," he said, "so I'd prefer the south end. I'll fire off once you've undocked," he said.

The platform could reach maybe eight miles an hour.

"Do you ever evac?"

"Sometimes. Rarely. Everything I have is here."

"My boat's pretty speedy," she said after a moment. "Why don't you come along? Take a few days off. Put a virtual sign on the door. Wait out the storm, come back, fix up. Bring the cat. Hell, bring the chicken."

She held her breath.

"I've got to stay with the bar," he said at last, and looked away. "Remote pilot doesn't stand a chance in that storm without me. Thank you, though."

She shrugged. "No thing." They locked eyes again and she kissed him quickly on the cheek and squeezed his hand, then headed out into the outskirts of the storm.

The sensors woke her for artifacts, and she corrected course, dozing in between. When the sky turned gray, she heated the last of her coffee. There was one seaweed farm offline, an hour away, but the weather was shit. She was tired and it would take all her fuel, but if she kept going, she could reach the islands.

In the late afternoon, she saw land. Harvesters and tankers were lined up to come in to the mega-biorefinery at Enue. The air stank of seaweed. Frigatebirds and boobies crowded the rocks. She scanned herself in at the small-craft marina at Bikini and made for her favorite of the island's five restaurants.

"Anything new on the menu?" she asked Lae.

"Vat-burger, homegrown, special sauce. Fresh tuna. Taro chips."

"Tuna and rice. With that yuzu sake spritzer thing."

At the counter was an Indonesian harvester with a pink hairband, nails green with seaweed. And the old man who practically lived here, whose bones must be the shape of that chair, Grumpy Jones. Between them was a tourist in a bright swimsuit and magenta muumuu, scribbling notes, asking the harvester about diving.

"History suckers," Jones said when the tourist left. "Coming to see drowned species. For the ghost reefs."

"Hardly ghost reefs," Vilma said. "The restoration crews have done an awesome job. I was here last month. Saw an octopus. Seahorses."

"Ghost reefs or the bomb museum," he muttered into his beer.

"Storm's passing north," observed the harvester. "Center's over Wake Island now."

"It looks vicious," Vilma said. "I had to rush to get out of its way."

"Hear about the ghost bar?" the woman asked. "Gone."

Vilma felt very cold.

"How do you know?"

"Distress call, went dark this morning. Heard it on the radio. Cyclone blew through its position. No one there to help out. Just empty water."

Vilma put down her drink and closed her eyes. She knew, she realized, almost nothing about him. *It's not fair*, she thought, and noticed she had spoken aloud.

The rain hammered for a long time, and they sat in its noise.

When she opened her eyes, Jones was opening a worn journal and drawing a black tally. The whole page, she saw, was filled with black lines. The harvester shook her head.

"What's that?" Vilma said. The dullness of her voice shocked her.

"He counts the people who've been killed," the harvester said. "By the carbon in the atmosphere."

"*They* had their fast boats and planes and frozen fish," Jones said. "And *we* end up dead."

"That seems pretty pointless," Vilma said, nodding at the journal. "You're hardly going to hear about all of them."

"He only counts the ones he has first- or second-degree connec-
tions with," the harvester said.

"Pointless," Vilma said again.

"He grew up on Bikini. Worked on records of losses when he was
younger. Not just numbers. Names, too."

When the rain let up at last, Vilma heard a squawk. "Look at that,"
she pointed. It was the largest bird she'd ever seen. "Looks like its
wings are dipped in black paint." Even the waitress turned to look.

"Gooney bird," the harvester said. "Sometimes they come around
nowadays."

"It's good luck," Grumpy Jones appraised.

Vilma raised her eyebrows. "Never thought I'd hear you say those
words."

"Everybody knows it's good luck."

They watched the albatross circle above the water.

"We could use some good luck," Vilma said at last.

"Lucky to even be here," muttered Grumpy Jones.

"Go on like this," said Vilma, "we're gonna have to start calling
you Optimist Jones. Make sure you put the bartender's name in your
book, too. Ghost bartender. For the record."

3

Regenerating

December, Montreal, −18°C / −0.4°F

The snow crunches under my feet as I make my way through a freshly gentrified red-light district toward the headquarters of the International Civil Aviation Organization. The ICAO building itself stands as a modern monument to global cooperation, bedecked with gifts from various member states. There are woolen tapestries depicting the birth of flight, a model of the world's first four-seat electric plane, a plaque from the United Arab Emirates that reads: "The civil aviation is an accumulation of human inventions, experiments, and cooperative efforts." I have entered this international workspace to join a few delegates to the UN Convention on Biological Diversity (CBD) for a workshop on geoengineering research. Many delegates are coming from the tropics, and most of us keep our coats on in the conference room, gazing out the windows as the steam from neighboring buildings rises and fades into the gray sky.

The delegates are members of a working group that gives scientific and technical advice to the CBD. They have been focusing largely on issues of synthetic biology, and on plotting pathways to a 2050 vision for biodiversity. Why tack on a geoengineering discussion to an already-busy week? As it turns out, the CBD is one of the only UN conventions to address geoengineering, largely due to the activism of environmental groups. In 2010, it issued Decision X/33, a statement that is often described as a moratorium on geoengineering. Later, the CBD took Decision XIII/14, which noted that more transdisciplinary research and knowledge sharing is needed to understood its impacts. This workshop was convened by an NGO focused on geoengineering governance to discuss what sort of research on geoengineering could be useful.

As snowflakes swirl outside, we watch a few expert presentations —a discussion of perturbed plumes and the chemical aging of substances in the stratosphere, the metaphor of geoengineering that grips "like opioids," the need to include instrumentation from other countries in geoengineering experiments. Then, we talk.

"What we thought was years down the road is facing us now," a delegate from a small nation declares. "My country has to be innovative." Another participant raises concerns that governance of geoengineering research is a bridge to development of further geo-engineering, warning against "engineering kinds of thinking." The earth is sacred and alive; the spoken or sung word is sacred. (These are not incompatible languages: UN negotiations take extreme care with words.) Pachamama has secret and sacred things, things you don't see. *We believe in correction and validation in the field with affection.* There are still some ways of seeing that are hard to reconcile; bitter histories of exploitation and damage that get mixed in and bleed together. *The rain, when it came down on our faces, was like lemons because of the geoengineering experiments.* They boil down to the same truths, though, in the end. *Our island is gone—who could imagine that would have happened? We will be extinguished.*

A good part of the discussion centers on responsibility. There's the sense, in some quarters, of the inevitability of carbon removal. Whose responsibility is the removal? A high-level official reported on a conversation in which a diplomat from a developing country about carbon removal had remarked, "That's for developed countries." More questions: about capacity, about doing the work and following through. Will there be moral hazards on a national level —that is, will geoengineering make developing countries less likely to keep forests? *With the storage—how are you going to contain that [carbon]? You want to remove and store, but you're also releasing?* A delegate from Africa asks: Do countries even have the knowledge to make use of or interpret the scientific results? Someone else asks: What kind of capacities, specifically, do we need? Another points out a problem with leaving governance and ethics up to individual researchers: when researchers in repressive regimes are told to do things, they do them. Researchers are workers and laborers, also working within systems with varying degrees of implicit coercion. A sudden noise erupts from my neighbors' bag, like a wildcat mated with a 1950s rotary phone and gave birth to a ringtone. Next topic: we're back to responsibility.

Polluters—emitters—must put the pollution back, it was forcefully asserted. *They must return it to the ground in natural ways.* But if it came from 3,000 meters below the ground in unnatural ways, asks an engineer, how can you put it back in a natural way? My discussion group gets stuck on a key topic of the day: natural climate solutions. These practices are a climate-focused subset of "nature-based solutions" or "ecosystem-based solutions," and are now an established part of the vocabulary in settings like these.

Understanding natural climate solutions

Natural climate solutions are conservation, restoration, and land management actions that either increase carbon storage or avoid greenhouse gas emissions from ecosystems. They're not considered to be geoengineering, generally, because they're often designed in conversation with communities and organized at local and regional scales. They're also focused on *mitigation*, and on avoiding further loss of carbon. Yet there's a potential tension within natural climate solutions, in that the activities or practices within ecosystem care or management range on a broad spectrum from highly interventionist to hardly interventionist at all. On the interventionist end, they can bleed into ideas of geoengineering—especially when they're conceptualized with the goal of removing carbon at a global scale.

"Natural climate solutions" is a fast-growing keyword. The language and framing around natural climate solutions has been developed over the past decade, with influence from conservation-based

NGOs like the International Union for Conservation of Nature, the World Wide Fund for Nature, and The Nature Conservancy, as well as high-level discussion by intergovernmental agencies like the UN Development Program and the European Commission, and the CBD. But grassroots environmental and agriculture-oriented groups are also excited about natural climate solutions, perhaps due to a desire for a better relationship with nature. Geographers Matthew Kearnes and Lauren Rickards depict new processes of (carbon) burial as a therapeutic relationship between humans and the underground, in a kind of "mirror image to the extractive processes of mining, drilling, and hydraulic fracturing."[1] Burial is reimagined as a part of earthly savior or redemptive processes. This redemption or restoration narrative emerges in tandem with new soil or ecosystem science (and in some cases, in contrast to it). As later encounters in this book will illustrate, this restoration narrative—which Kearnes and Rickards describe in conjunction with a "land aesthetic" that is deeply *moral*—is a powerful force underlying the rise of natural climate solutions, and terrestrial carbon removal in particular.

Yet at this Montreal geoengineering research workshop, many participants expressed a feeling that these natural climate solutions were not being given their due in the big Intergovenmental Panel on Climate Change (IPCC) reports, or in the international climate policy the reports inform. One official noted that only 2.5 percent of global climate finance goes to ecosystem-based approaches. These two UN conventions, the UN Framework Convention on Climate Change (UNFCCC) and the CBD, participants reported, were likened to two different planets—the same parties participate in both conventions, but they have totally different stories when it comes to the climate issue. The institutional competition between the two conventions reminded me of organisms seeking food—only in this case they sought attention, respect, and funding. Institutional politics aside, the discussion broached two questions: Were natural climate solutions in fact being ignored in global climate politics? Moreover, how useful can they be?

In this chapter, we'll look at some key nature-based approaches for carbon dioxide removal—regenerative agriculture, afforestation, soil carbon sequestration, biochar, and blue carbon—in an attempt to sort out the discourse from the facts in the field, and examine what it would take to make these solutions part of our climate future.

Carbon farming: Traveling the regenerative spiral

"Eat pizza. Drink beer. Save the planet." This virtual flyer appeared on a soil health Instagram account, luring me to Café Gratitude on Rose Avenue in Venice, California. They serve pizza and beer made with Kernza, a Eurasian forage grass that's been optimized for carbon storage. Foamy and rich, the Long Root Ale lived up to the description on the can: the Kernza added "a slight spiciness to the dry, crisp finish." According to the can, "Kernza also pulls carbon out of the atmosphere and stores it in the deep roots and in the soil. You don't get carbon credits but it's a damn good beer."

Kernza is a perennial grass, so it doesn't have to be replanted every year and can be grown without tilling, which means it sequesters more carbon. The roots extend ten feet deep, or twice as deep as wheat, which is key for soil carbon sequestration. Kernza (a trademark) has been studied since 1983 by plant researchers at the Rodale Institute, who worked with US Department of Agriculture researchers to select for fertility and seed size.[2] Kernza was further improved and selected by a Kansas-based nonprofit, the Land Institute, who hopes to ultimately develop a variety with yields similar to annual wheat. With advances in gene sequencing, which helps them select which plants might offer desirable traits, it's been possible to rapidly improve the variety (without genetic engineering). The agribusiness giant General Mills gave the University of Minnesota half a million dollars to work with the Land Institute to study it. This particular beer actually emerged through Patagonia Provisions, the food division of the outerwear company, who funded the initial steps to get it to market.

At first glance, this story—I walked into a café in Venice and bought a beer—might read as a dismissible tale of empty green consumerism. I mean, the pizza came served on a plate that read "What are you grateful for?" in Bambino font. However, dismissing it offhand would occlude the several decades' worth of effort a remarkable constellation of actors put in so that this Kernza beer could reach my table. It took interest from everyone: government and university researchers to nonprofits, big agribusiness, environmentally edgy corporations, and the grassroots community advocates that drew my attention to the restaurant. All of them bore some degree of devotion to a vision of regenerative, carbon-sequestering food systems.

"Regenerative," etymologically, comes from "giving birth again." It was first used in a design context by Buckminster Fuller, and "regenerative agriculture" was coined by publisher and organic advocate Robert Rodale in the 1980s.[3] "Regenerative agriculture" refers to methods that manage land holistically for carbon sequestration, crop resilience, soil health, and nutrient density. Again, though, this type of farming is more than a collection of agricultural practices—regeneration is a wider narrative linking sustainable business with agrarian culture. To understand all this better, I talked with Finian Makepeace, a musician and soil health enthusiast who leads speaker trainings for an organization called Kiss the Ground. The trainings equip soil health advocates to go out and talk with people about why soil is important. Clearly, Makepeace has thought a lot about how to spread the message.

"I used to present to people about regenerative ... and I had a constant feeling that people weren't really understanding what I was saying." He explains that it's not that we can't understand what "regenerative" means. It's that we've lumped it into our preexisting container of things that are "environmental." "And unfortunately our preexisting container ... to put it bluntly, [is] not as epic as the regenerative container." Our preexisting container is the "sustainable" container: the impulse to "do less harm, stop messing up the earth, stop taking too much." The recycling symbol, with the arrows forming a circle, is emblematic of this sustainable container. "That recycling symbol is very easy to recognize. And most people will say, 'Yeah. That's where I'm at.'"

We need to think outside the circle, Makepeace insists, and expand our thinking beyond mere sustainability. To be "sustainable" is more like rearranging the deck chairs on the *Titanic*. It doesn't do enough to move us beyond an extractive, degenerate relationship with nature. A regenerative approach is thus not about doing less harm, but about healing and restoration. His way of explaining it is with a regenerative spiral. Regeneration goes beyond just letting nature recuperate; it's about actively working to increase flourishing.

"Carbon farming" is another term that's erupted into the sustainable food movement, and it has similar aims as regenerative agriculture, though the term is narrower and more specifically focused upon storing carbon in agroecosystems. Journalist Michael Pollan called carbon farming agriculture's secret weapon; environmentalist Paul Hawken called it "the foundation of the future of civilization."[4] Carbon farming advocates emphasize the role of agriculture in contributing to climate change. Conventional agriculture is presently a massive source of emissions, with land use change contributing 25 percent of total anthropogenic greenhouse gas emissions (10 to 14 percent from agriculture; 12 to 17 percent from land cover change).[5] An increase in soil carbon is accomplished in these key ways: (1) switching to low-till or no-till practices, (2) using cover crops and leaving crop residues to decay, and (3) using species or varieties with greater root mass. Double-cropping systems, where a second crop is grown after a food or feed crop, keep more carbon in the soil, as well. Much attention has also focused on multistrata agroforestry systems, including edible forests or food forests, and silvopasture, which involves grazing animals below trees.

Regenerative grazing is another version of regenerative agriculture. The idea is that grazing animals, such as cattle, are managed to mimic how animals grazed on the grasslands in pastoralist societies. They eat everything in one area as a herd, till the soil with their hooves, and then move on to another area, allowing the first to regrow. The practice is promoted by Allan Savory, a biologist from Zimbabwe who heads the for-profit Savory Institute. Savory's TED talk, "How To Green the World's Deserts and Reverse Climate Change," made its debut in 2013 and is on its way to 3 million

YouTube views. Yet regenerative grazing has faced skepticism. As one critique in a peer-reviewed journal stated, "The false sense of hope created by his promises, expressly regarding some of the most desperate communities, are especially troubling. Scientific evidence unmistakably demonstrates the inability of Mr Savory's grazing method to reverse rangeland degradation or climate change, and it strongly suggests that it might actually accelerate these processes." The scientists noted that rangelands are weak carbon sinks because plant production is water limited, and that the ecological benefits of "hoof action" to rangeland restoration were overstated.[6] But Savory's method has scores of passionate advocates. They believe that mainstream science is too reductionist to see the potential. He's become a folk hero, and something of a counterpoint to modern, corporate-sponsored science. Geographer Rebecca Lave has described his work as an example of "free-range science": "low-budget, informal, strongly regional, and without the trappings of professionalized laboratories and tools." In this form of knowledge production, she writes, scientific authority stems from market take-up rather than exclusively from academic prestige or peer review.[7]

"Allan Savory goes without shoes to pick up subliminal information about the land that he walks," report L. Hunter Lovins and colleagues in *A Finer Future: Creating an Economy in Service to Life* (2018). The authors suggest that regenerative agriculture is "the one real shot we have to counter the climate crisis," citing several sources who "believe that regenerative agriculture can displace all of the carbon emitted by humans each year and begin rapidly reversing global warming."[8] Indeed, it comes down to a matter of belief; there are a lot of bold claims about what tending the soil can accomplish. This book is not an outlier here; rather, it is one example of an emerging genre. If this potential is genuine, why is it being ignored? Another book, Charles Eisenstein's *Climate: A New Story* (2018), explains that regenerative agriculture remains marginal despite its vast potential because it is incompatible with conventional regimes of measurement. Its dynamic and locally tailored practices are incompatible with scientific protocols, meaning that, much like holistic medicine, it can't be studied.[9] The data remains anecdotal, rather than quantitative, he notes, and so it can't be translated into policy. Ultimately, Eisenstein writes, "we are being invited into a different way of engaging the world ... A civilization that sees the world as alive will learn to bring other kinds of information into its choices."[10]

But what *does* the peer-reviewed science say about the potential of soil carbon sequestration approaches? The principles behind soil carbon sequestration are sound and fairly well understood. To grasp its potential, we have to understand the depleted state of soils today. Soils are vast reservoirs of carbon: they hold three times the amount of CO_2 currently in the atmosphere, or almost four times the amount held in living matter. But over the last 10,000 years, agriculture and land conversion has decreased soil carbon globally by 840 gigatons, and many cultivated soils have lost 50 to 70 percent of their original organic carbon. Intensive crop cultivation can reduce soil carbon by 25 to 50 percent in just thirty to fifty years.[11] The good news is that this can be reversed. The soil carbon sequestration initiative "4 per 1000," announced at the 2015 United Nations Climate Change Conference in Paris, has the goal of increasing soil concentration 0.4 percent per year, which would increase the carbon sink by about 4.3 gigatons carbon dioxide equivalent (roughly equal to the emissions of a large emitter like the European Union). One policy brief points out that the 4 per 1000 initiative would cost $500 billion per year, which is in the same ballpark as current agricultural subsidies worldwide.[12] New technological capacities could also play a role in reaching these goals—for example, designing crops to have root architectures that could store more carbon.[13] Scientists at the Salk Institute, one of those La Jolla–based biotech research institutions, are developing an "Ideal Plant" with carbon removal traits that can be switched on either via traditional breeding or via the gene editing technology CRISPR. At the heart of their strategy is the effort to induce plants to produce more of a molecule called suberin, a biopolymer that's the main component of cork, which could help roots resist decomposition and store more soil carbon.[14]

However it is done, removing four gigatons of carbon dioxide a year via agriculture would be ambitious and fantastic. Even then, it's only a tenth of what we're currently emitting. Still, every bit helps.

It's important to understand, however, that soil carbon accrual rates *decrease* as stocks reach a new equilibrium.[15] Sequestration follows a curve: the new practices sequester a lot of carbon at first, for the first two decades or so, but this diminishes over time toward a new plateau. Soil carbon sequestration is therefore a one-off method of carbon removal. When the potential is used up, this is called "sink saturation." It's also reversible, meaning the new practices must be continued to keep the carbon sequestered.[16] And so, if negative emission technologies are expected to be needed later in the century—but we started these methods now—the sinks would already be saturated by mid century.[17] Calculations of yearly potential elide this fundamental aspect of the soil carbon contribution. On the other hand, regenerative farmers would argue that you actually *gain* from implementing these practices, and that since the transition costs are up front, there's no reason why farmers would want to stop them once they've made the transition. What's more, these are no-regrets solutions, as they simultaneously improve soil quality.

The eventual climate restoration potential may be the wrong place to focus on right now. Kristin Ohlson is a writer whose book *The Soil Will Save Us* ranges through several farmlands, from North Dakota to Zimbabwe. She deftly addresses this issue of quantifying how much carbon is actually sequestered.[18] As she notes, it's a tough undertaking: there's not enough research. "There certainly is a problem with ag departments and ag schools being heavily, heavily funded by businesses that have a stake in the status quo, have a stake in the kind of agriculture that uses a lot of chemicals, does a lot of tillage, and requires a lot of equipment and all of that," she tells me. The soil is a complex system, and it's hard to pull out one factor and change it, and then compare it to a control.

"The soil health practitioners see such dramatic changes on their own land because they're doing a lot of different things. They're doing cover crops, and they're doing compost, and they're doing no-till, and they're bringing in their animals to eat down the remaining vegetation. They're doing all these things that build up life in the soil and carbon in the soil, so I think it's natural that those people would feel very impatient with the slow and reductive pace of university science." As she says, the soils do become carbon rich. Ohlson points to the benefits to the whole system: reduction in runoff of fertilizer and other chemicals that pollute waterways; decreased air pollution from blowing dust; and increased soil permeability, meaning the system can resist droughts and floods. "There are so many benefits that come with this," she adds, "that the average person, even if they don't have a number to hang onto for carbon storage, for carbon sequestration ... should still support this shift—which is a paradigm shift in agriculture, because it has so many benefits."

This may be a challenge for those of us fixated on empirical data or climate change narrowly defined, but it's a useful way of looking at reasons for pursuing soil carbon sequestration. Noah Deich, executive director of the nonprofit Carbon180, also emphasizes the importance of moving in the right direction; the need right now is to begin the work, rather than get caught up in extra-precise quantification. "The question might be a question less of what is the ultimate scale potential, but what are the incentives? What are the policy designs? What are the corporate action campaigns? What are the consumer engagement campaigns that can reward producers for managing their land in a way that sequesters carbon on them? That's the first step ... we need to just get started," he says.

Biochar

Over 2,000 years ago, in the Amazon, indigenous people were managing soils to be carbon rich. Deposits of these dark, fertile soils, called *terra preta*, can still be found today. *Terra preta* soils have inspired many advocates for biochar, which is essentially carbonized organic material that benefits soils by making them more fertile and helping them retain water. The basic idea is that biomass—crop residues, grass, other plants or trees—is combusted at low temperatures (300 to 600°C) without oxygen. This process (pyrolysis) results in charcoal, which is a form of organic carbon that can endure.

Permaculturist Albert Bates got turned on to biochar during his travels in the Amazon. Bates is an environmental rights lawyer, a cofounder of the Global Ecovillage Network, and a long-term resident at The Farm, an intentional community in Tennessee. He's been concerned about runaway climate change since the 1980s, and in 1989 he wrote his first book on the climate crisis. "I've been searching for solutions," he tells me over Skype from Mexico. "I was kind of despairing over everything, until I went to Brazil for a permaculture course ... and got to travel into the Amazon and visit with some scientists there, and study the *terra preta* soils, and bring some back to Tennessee."

In their book, *Burn: Using Fire to Cool the Earth*, Bates and fellow biochar researcher Kathleen Draper describe various ways that biochar can be used to store carbon. They sketch out how we can go from wasting to banking carbon in virtuous cycles they call "carbon cascades." The authors travel through the world of biochar projects, from a village-scale biorefinery in China to an eco-lodge in the Dominican Republic. One option for carbon removal is what Bates and Draper describe as bioenergy with biochar capture and storage (or BEBCS, in contrast to BECCS). In our conversation, Bates explains to me that there's more that can be done with biochar besides enriching soils. Biochar can also go into road construction materials, and into aggregates used in cement and concrete. "Turns out that if you elevate the content in concrete with 8 percent to 12 percent biochar, it actually improves the quality of the concrete over what it had been with just sand," Bates tells me. It can also be used to improve permeability of surfaces, as has been experimented with in Stockholm. "We've got roads; we can look at the bitumen in asphalt. We can look at bridges, airports. We can start to think about composites other than steel, concrete and asphalt; we can think about plastics, and the monomers and polymers that go into the plastics. Many of those are enhanced, it turns out, by carbonates. So you can add biochar, pyrolized carbon, into composites, and now you get a stronger polymer."

Using biochar in the built environment would make it possible to employ biomass energy without being reliant on forestry or biomass energy crops, because it would be possible to use things that wouldn't normally go into agricultural soils. "When you start to talk about putting it in cement, putting it in highways, putting it in airports and roads and things like that, bitumen—now you can stand to add in plastics from municipal waste dumps and bio-solids from sewage treatment plants." Expanding the feedstock possibilities in this way would be a real breakthrough. "You don't divert from food or from biodiversity services of forests in order to feed your biomass energies," he tells me. "You can get your energy from pyrolysis from your sewage treatment plants, from your municipal landfills. All of those can go to make energy for you in vast quantities."

In *Burn*, Bates and Draper look at the potential of biochar when one includes sources like municipal waste, landfills, and so on. With biochar in agricultural soils, Bates comments that the drawdown potential would be one to four gigatons of carbon dioxide equivalent (or, similar to that of soil carbon), when emissions are around forty gigatons. "It isn't enough. We can talk about tree planting, we can talk about seaweed, we can talk about kelp farming … You sum them all together, and you're lucky if you get to seven or ten or even twelve gigatons of CO_2 per year removal." But if you look at biochars in these new sinks, "concrete airports and buildings, carbon fiber cars, polymers of various kinds, it works out to closer to fifty gigatons."

And it could scale very quickly, Bates argues: "Now we can be talking drawdowns … with ten gigatons a year coming out of the atmosphere, you can begin to calculate how many parts per million we can go down. From 410 down to 400 down to 390 down to 380, 370 and so forth. Well, yes and no. There's some problems with that approach to this. We've been adding carbon to the atmosphere for 100 or 200 years. We know that carbon added to the atmosphere makes it warm up; we know how the greenhouse effect works. We don't know how fast it responds when we start to withdraw carbon. We deprive the atmosphere of photosynthetic carbon: How fast does the temperature respond? How fast does the chemistry of the atmosphere and the oceans respond? We don't have any data set for the reverse of warming." So, Bates cautions, we don't actually have any proof of this. "That stands as a theory. That's waiting to be tested."

There isn't much peer-reviewed research on applications of biochar in the built environment, and virtually none of the scholarly meetings I attend on negative emissions science or governance discuss it. I ask Bates why it is that a writer and lawyer, a self-described hippie who lives in an intentional community and crowdfunds his work on Patreon, is the one who's looking into this. Bates, though, actually seems quite optimistic about the prospect that really good research universities and institutes will come in and up their game. Part of the promise of this idea, Bates points out, is that it flips the problem from just being about carbon to being about waste management. Another benefit is that it wouldn't require all the injection of carbon that CCS projects entail. "We don't need to gasify or liquefy carbon and pump it a mile under the earth or a mile down into the ocean, which would be a bad idea for a number of reasons," he explains. "Instead we can just solidify it into hardscape and build our cities of the future, roadways and things of the future, that way." Rethinking our relationship with carbon and our view of waste is a beautiful vision. Their book is filled with frontline journalism documenting the creativity on the carbon frontier, where interesting ideas receive all too little attention from establishment science and policy, and hope blooms in unexpected places like sewage sludge or sidewalks.

Planting forests

Planting trees may be the most beloved option for dealing with climate change. There's a beautiful kindergarten simplicity to the image. Planting a tree is touted as something that every community can do, and communities benefit locally from the psychological and climate benefits of the green space. Reforestation on climate-significant scales, though, is a different beast.

One government that has been thinking about forest carbon sinks on a nation-state scale is Bhutan, the world's first carbon-negative country. They manage this feat in part because of an abundance of hydropower, and because the country's forests suck up more carbon than the nation produces. Bhutan's constitution, enacted in 2008, mandates that a minimum of 60 percent of the country's total land area must be forested at all times. It's not just pure luck of geography, then, that keeps the country carbon negative; it's also governance. What enables that kind of leadership?

The year the forest-protecting constitution was enacted was the same year that the new king, a twenty-eight-year-old reformer, was coronated. This was no ordinary coronation: the elder king had announced the transition of the monarchy into a democracy. People had come by truck, motorbike and yak from all regions of the country to see the former king place the Raven Crown on the head of his son. The ceremony was held on November 6, 2008, a date deemed auspicious by three enlightened astrologers. A Thursday: the eighth day of the ninth month of the earth male rat year.

I was visiting the country then, and sharing in the nation's jubilant mood. The day before, Obama had claimed the US election. I was glued to the blurry BBC coverage, guiltily monopolizing the television of the Bhutanese family whose farmhouse I was staying in. Occasionally, an eighteen-year-old boy named Singye would join me in watching the coverage. We sat upon carpets in the bare room and drank tea while I tried to explain what was going on with all the revelers in the streets of Chicago: a quarter of a million Americans, or more than a third of Bhutan's population, turned out to celebrate.

"We have a new leader right now, too."

"What's his name?"

"Barack Obama."

"Is he married?"

"Yes, with two children." Singye nodded.

In Bhutan's capital, Thimphu, they celebrated their new leader for three days: solemn ceremonies, concerts with traditional dance, archery. In honor of the coronation, all mobile phone communications were shut down during the day. The streets were lined with Buddhist flags: bright red, yellow, blue, green, white. At night, the trees and buildings were festooned with strings of rainbow light.

In my conversations, I had been surprised to learn that people were excited about the transition to the young Dragon King, but often less enthused about the transition to democracy. People adored the monarchy. They'd seen conflict and corruption while watching elections in neighboring India, and they weren't sure if they wanted their country to go that route. I observed the festivities from a hill, across the river from the seventeenth-century fortress in Thimphu where the coronation ceremony was taking place. While watching the crowds gather, I ran into another foreigner—a development professional. I told him what I'd been hearing about this governmental transition, about the wariness toward democracy. He responded with an anecdote about working in Vietnam, a one-party state that could implement policy shifts quickly. One day, the government had decided that everyone needed helmets for motorcycles, and the next day, everyone had them. "Like that," he said, snapping his fingers.

Bhutan is a standout example of how even in a democracy, a carbon-negative target can be achieved. It doesn't necessarily require autocratic fiat or loving decree to make these land-use decisions. Of course, the practicalities of scaling up forest management and carbon-negative land use beyond a small, mountainous, sparsely developed country prompt the question: What kind of government could achieve this goal? To sequester one gigaton of carbon dioxide, one would have to afforest 70 to 90 million hectares, or a land area about twice the size of California.[19] Now, again, current emissions are forty gigatons of carbon dioxide. So that's a tremendous effort just to put away one gigaton. A few gigatons are certainly doable, since there's a lot of farmland that's been abandoned because of low productivity; one conservative estimate put the acreage at around the size of India.[20] Scientists calculate that there are large areas of land available for reforestation—from 500 million hectares in bottom-up estimates, to between 1 and 3 billion hectares in modeled estimates of nonagricultural land broadly.[21]

But how can the transition be orchestrated, and how many times can that feat be repeated? A finite number of times. Consider figures like those outlined in "Alternative Pathways to the 1.5°C Target Reduce the Need for Negative Emission Technologies," a helpful analysis by a European group of modelers (reported eagerly in the environmental press as "World Can Limit Warming to 1.5°C 'without BECCS.'")[22] How did the modelers do it? For starters, they assumed capture of 400 gigatons by reforestation. But here's the assumption you have to dig out of the details: in their scenario, agricultural efficiency increases, and massive areas of cropland and grazing land are converted to forests. Next, this storyline "assumes a technological breakthrough and mainstream acceptance of cultured meat, starting in 2035 … We assume that by 2050, 80 percent of meat and eggs (but not fish and seafood) are replaced by cultivated meat, which is grown directly from corn and small amounts of soy."[23]

Indeed, if we switch to "cultivated meat," grown from cells in vats, or just to plant-based meat, then we can plant trees on the vast swaths of land that are currently being grazed for meat production. But at this point, we're not just talking about an afforestation project. We're talking about cultural and behavioral change in countries who value meat, including getting them to accept something as new and potentially weird as lab-grown meat; about telling people who just gained the economic capacity to *have* meat that they should switch to something else; about defanging a powerful industry lobby; about telling scores of pastoralists that they need to adopt different livelihoods. This is more than "afforestation" in a simple sense. It's a social project, enrolling education, public health, and more. Afforestation on this scale is basically geoengineering. Whether that dramatic transformation is easier, better, or more desirable than all the other approaches to removing carbon should be a vibrant matter of debate. But it's not as simple as planting a seed in the earth.

Talk of forest creation begs the question: What exactly is a "forest"? There are things that look like forests, but they may not be carbon rich. A recent study in *Science* showed that Europe has been considerably afforested since 1750—with an increase of 10 percent over this period (most of it from 1850 to present). During this transition, 85 percent of the forests were put into management. Yet two and a half centuries of managing these forests has not contributed to cooling. Instead, converting deciduous forests into coniferous forests changed the albedo (the proportion of the light reflected by the earth's surface), canopy roughness, and evapotranspiration from the land, which warmed things up. Europe's forests accumulated a carbon debt of 3.1 gigatons over this time, as the extraction of wood released carbon that was stored in biomass, litter, deadwood, and soil carbon.[24] Most biomass carbon is in the woody stems and roots of old trees, and primary forests store 30 to 70 percent more carbon than commercially logged or plantation forests. It takes hundreds of years to grow these carbon stocks to their natural capacity.[25] There's also a big difference between tropical and boreal forests, and the net climate effect of increasing boreal forest is unclear.[26] Moreover, scientists have been sobered to find that trees can emit methane and volatile organic compounds, which could offset their cooling effects.[27] Complicating it further, afforestation schemes need to be "climate smart" by accounting for projected climate impacts, including extreme storm events or outbreaks of forest diseases and pests. Forests may be a risky place to bet on for carbon storage, because in the event of wildfires or die-offs, the carbon could be lost.

Yet from Bhutan to the Sahel's so-called "Great Green Wall" to China's national reforestation project, aspirations for afforestation and reforestation are vast. Emerging technologies, like aerial planting of seedlings via drones, may aid states and organizations in their efforts. Theoretically, governments can play a large role in this. Much forestland is within the purview of governments: one-third of Latin American forests, about two-thirds in Asia, and virtually the entire area of forests in Africa.[28] However, one problem with making calculations around the ability of countries to command afforestation or reforestation is the assumption that developing countries have full control over the lands and actors within their borders, notes geographer Jon Unruh. He points to the problems of enforcement, deep and long-lasting resistance to and suspicion of land-related policies, corruption, and discrimination.[29] Existing forest carbon schemes such as REDD+ (Reducing Emissions from Deforestation and Forest Degradation, the program under the UN Framework Convention on Climate Change) have run into a host of documented problems: the inadequacy of certification schemes to protect livelihoods and biodiversity when pursuing climate goals,[30] the ways in which "carbon colonialism" via plantation forestry amounts to neoliberal land grabs,[31] and many others. On the other hand, research has also shown how agricultural intensification, land use zoning, forest protection, increased reliance on imported food and wood products, and foreign capital investments can all work together in managing land use transitions.[32] What's clear is that it's not useful to treat afforestation as something that happens "over there," in the forest. Rather, it's a complex social project that touches all of us—at the very least, through what we choose to eat every day. That juicy hamburger could instead be a tree storing carbon: modeled pathways for keeping warming to 1.5°C assume that it will become one.

Blue Carbon

Carbon stored in peatlands, mangroves, tidal marshes, and seagrasses is collectively known as "blue carbon." These areas are thought to be hot spots for storing carbon, and so one of the best things for the climate would be to stop destroying wetlands. One-third of the world's mangrove, seagrass, and salt marsh areas have been decimated over the past several decades.[33] They are being degraded at devastating rates—in some instances, up to four times that of rainforests.[34] Between 2 and 7 percent of blue carbon sinks are being lost annually, which is a crazy rate of decline. Protecting these ecosystems could contribute powerfully to mitigation. One UN report estimated that doing so could amount to 10 percent of the reductions needed to keep CO_2 concentrations below 450 parts per million.[35] Seagrass meadows are particularly impressive: they can sequester carbon for millennia. Sediments in healthy coastal ecosystems can continue to accrete carbon vertically as sea levels rise. This means they can keep building up carbon, unlike the terrestrial carbon sinks, which become saturated in a few decades.

Enhancement of blue carbon via wetland restoration and protection seems like one of the carbon removal approaches with the fewest drawbacks. It's something that can go alongside existing restoration and coastal adaptation / shoreline protection projects. For example, biochar and other carbon-rich materials could be used in these projects to sequester even more carbon. Despite its potential importance, blue carbon has scarcely been addressed in the literature on "climate engineering," at least up until the 2018 National Academies report put it on the research agenda for "negative emissions." I asked marine geochemist Sophia Johannessen about this apparent omission. She explained that it's a new field: "These papers started to appear in about 2001, and the field has expanded rapidly since them." Johannessen is actually at the center of a lively scientific debate on this topic, in the wake of a 2016 paper she published with fellow Fisheries and Oceans Canada colleague Robie Macdonald: "Geoengineering with Seagrasses: Is Credit Due where Credit Is Given?"[36] In it, they argue that while seagrasses are reported to account for up to 18 percent of the carbon burial in the world's oceans, the accounting is wrong because it doesn't address how carbon is deposited in marine sediment. In fact, estimates may be off by 11- to 3,000-fold. I asked her what accounts for the incredible variance in assessments, and she speculated that it might be disciplinary boundaries between two communities of research. "The people who are publishing these papers, saying that there's a huge sequestration potential in seagrasses, haven't been working in marine sediment geochemistry," she replied. "They know about the biology of seagrasses, but they don't really understand how sediments process and sequester carbon."

Biologists are naturally intrigued by sequestration in seagrass meadows because seagrasses are crucial habitat for juvenile fish. But these ecosystems, being right at the coastlines, are under threat from urban development. Carbon sequestration would add a compelling reason to protect seagrasses. Yet to a marine geochemist, carbon sequestration in seagrass seems elusive. First of all, most of the carbon that's sequestered in coastal sediments isn't sequestered where seagrass grows. Organic carbon sticks best to very fine particles. But seagrasses generally grow in coarser sediment, like sand —so the places where the seagrass is growing aren't really the places where carbon tends to build up. Another issue is that a lot of global estimates of carbon sequestration in seagrass meadows are based on measurements of one specific type of seagrass that grows in beds in the Mediterranean and Southern Australia, *Posidonia*. It has extremely long root mattes that extend for meters into the ocean floor. These root mattes are not the norm for seagrasses generally, yet measurements from *Posidonia* haven been extrapolated throughout the world.

Coastal sediments, root mattes: this all sounds pretty obscure. But the stakes for getting it right are high, because the first international protocols are now emerging around the voluntary market for carbon burial via seagrass. Sequestration in seagrass meadows could therefore be used to offset emissions elsewhere. "But if the seagrasses really aren't actually storing as much as people think, and credits for planting seagrasses are used to offset emissions elsewhere, then the net effect could be an increase of carbon emissions into the atmosphere—just the opposite of what carbon credits are supposed to achieve. So it's a really important topic, even though it sounds boringly technical." Johannessen explains, summarizing the short story: Seagrasses are important for habitat and protecting coastlines. "Some people have said that seagrasses also bury a lot of organic carbon, and that we should be able to claim carbon credits for protecting or restoring them. Marine geochemists reply that the current accounting methods are wrong, and the estimates are far too high. But there is methodology that can be used so that people actually could assess carbon storage in seagrass meadows properly, and there is a potential for it to be used for claiming carbon credits in the future."

On one hand, it's disheartening to see how information about carbon removal can fall into this void between disciplines and lead to these technical mistakes in calculating global potential—we could probably find similar illustrative examples for every other carbon removal technique, too. On the other hand, it's promising to see the scientific process engaged and self-correcting, lending hope to the idea that maybe we will have better estimates to guide us in the future. Yet even if we did have perfect knowledge about how much carbon all these practices could sequester, would people act on it?

Will natural carbon removal fulfill our hopes?

These regenerative, land-based approaches to carbon sequestration are exciting in part because people bring such love, care, and devotion to them. Through these conversations about regeneration and natural climate solutions, people are developing a broader vision for how to live with the earth in the future.

This capacity to envision alternate futures is precisely what is needed. It is invigorating a stale climate politics with grassroots energy. And yet, in terms of the capacity of natural climate solutions to not only mitigate but *reverse* climate change, there are three sobering realities that must be kept in mind.

The first is the difference between one-off removal and continual removal. Both large-scale afforestation and soil carbon sequestration face this issue of being techniques whose storage capacity levels off over time. Second, many natural climate solutions aren't permanent—they require maintenance and could be reversed by future decisions, as well as by climate change itself. For example, some research on conservation tillage shows that it may only increase soil carbon to a depth of ten centimeters. Because this thin layer is very vulnerable to changes in management, agricultural practices may not sustain surface soil carbon over the centuries or millennia for which climate policies must be designed.[37] However, this maintenance requirement should not be thought of simply as a burden or liability. "Though premised on a logic of escape and exit, the marking of buried materials is also tied to the marking of sites of ongoing obligation," write Kearnes and Rickards.[38] These types of obligation and care are of a very different logic than our society has been running on. But care and the intention to maintain carbon can only go so far: if left unchecked, climate change could reverse terrestrial carbon sinks by mid century.[39] Wind and rain will erode soil into the ocean; fires will burn down forests. In this sense, the efficacy of the solutions is vulnerable to the problem itself.

A third reality concerns the scale of these measures' results compared with the scale of emissions. Remember, current emissions are on the order of forty gigatons of CO_2 per year, or fifty gigatons of

CO_2 equivalent when you count other greenhouse gases. Afforestation, soil carbon, and biochar, at the *extremes* of their socio-technical potential, could remove perhaps ten to twenty gigatons of CO_2 equivalent per year of that (as per the 2017 UN Environment Programme's *Emissions Gap Report*). But this would be a tremendous work of social and industrial transformation. This highlights, first, the need to stop emitting. It's impossible for these techniques to make up for our current levels of carbon emissions. And this peak in emissions needs to happen right away. For even if you view the problem on a longer time frame, say 200 or 300 years, those natural sinks will be sequestering their ten gigatons CO_2 equivalent per year for, say, 50 years—then they will be full.

Try this thought experiment. Imagine that emissions flatline in 2020; the world puts in a strong effort to hold them steady, but it doesn't manage to start decreasing them until 2030. It's plausible that it would take ten years to start a worldwide decrease, right? But ten years steady at 50 Gt CO_2eq—and there goes another 500 $GtCO_2$eq into the atmosphere. That one decade would cancel out the 500 $GtCO_2$eq the soils and forests could sequester over the next fifty years (sequestered at an *extreme* amount of effort and coordination among people around the whole world). Plateauing emissions for a decade before starting to bring them down might not sound that bad—the world would probably celebrate that triumph—but it would take fifty years of Herculean effort in enhancing these natural sinks *just to make up for that decade alone* of flat emissions. Then, the sinks would be maxed out in terms of additional removals. But we'd still have to maintain the storage, or we'd risk releasing the carbon again. Moreover, there wouldn't be enough sink capacity later on in the century to make up for small amounts of continued emissions from some hard-to-mitigate sectors, like shipping, aviation, steel, or rice production.

The above calculations, rough as they are, lead me to conclude that it's very risky to rely on natural climate solutions alone. I'm concerned we could risk placing all of our regeneratively grown eggs into one lovely, but small, regeneratively grown basket. If you're absolutely sure not only that emissions will decline very sharply over the next ten years, but that natural climate solutions will be as effective as we hope, and also that global demand for meat will stop rising and decrease dramatically—if you would bet it all on everything going just right—then okay, perhaps natural carbon removal is all we need. I don't know anyone that would place that bet, however. Moreover, I'm worried about what happens when books are published, countless YouTube videos are recorded, and conference halls are packed with people, all propagating the notion that soil can magically suck up all the carbon that's been burned. It's so intuitive that we should be able to save the planet through care: regenerative agriculture seems like it *should* be the right answer. It feels right and good to source hope there. Nevertheless, I'm concerned that this determined post-truth faith in soils could contribute to a failure to invest in other technologies that are also needed for this gargantuan carbon removal challenge.

When it comes to afforestation, regenerative agriculture, biochar, and blue carbon, climate change may actually be the least compelling reason to take many of these actions. We should regrow complex forests for their biodiversity benefits, establish agroforestry because it increases farmer resilience, recycle crop residues in order to create a circular economy, improve farm information systems to help farmers, and take care of the soil so as to avoid dead zones in the oceans—and much, much, more. Climate benefits are a fantastic bonus to these efforts, but the current magnitude of emissions overwhelms what these sinks can address during this century. We are deluding ourselves if we think this can be the *only* response to the disaster faced by people and species around the world. Natural climate solutions should be pursued with all the energy we can muster, and they really *can* make a contribution. But if we genuinely care about lessening climate impacts, curbing sea level rise, or saving species, other measures will also be needed.

Part II: Burial

4

Capturing

*Southeastern Saskatchewan, Canada,
summer, sunny, 77°F / 25°C*

The most interesting part of the power plant was, in fact, the tabby cat. It wound its way around the greenhouse, gingerly stepping among sprouts and purring warmly. The ghost presence trotted behind as I toured the greenhouse's chambers, jumping between the trays of seedlings and the seeding machines.

The greenhouse lay in the shadow of Shand: a hulking coal-fired plant that is also home to the Shand Carbon Capture Test Facility, which serves as an experimental area for clients to test new carbon capture equipment. It's owned by SaskPower, the state power monopoly of Saskatchewan. The waste heat from the coal plant warms the little trees through the winter. Over the past twenty years, Shand's greenhouse has handed out 10 million seedlings for free to people who are creating prairie shelterbelts or restoring habitat.

We are drawn to the lively—this striped cat, full of personality and intentions; the seedlings, aiming to grow. Infrastructure, on the other hand, strikes most of us as boring. Cold. Pipes, vents, bolts. Or hot, but inhuman, of another realm: inhospitable. Infrastructure is entangled three-dimensionally in landscapes like these. Picture Shand, rising between the coalfields with their draglines scraping. Then, fourteen kilometers due west of the Shand power plant and greenhouse, lies Boundary Dam Power Station—yet another coal-fired power plant. However, this one is the first generating station to use carbon capture and sequestration. Picture the Aquistore Project, a few kilometers west of that, shooting carbon 3.4 kilometers down, through a layer of brine-filled sandstone, into the Deadwood and Black Island formations. Picture the pipeline also snaking north toward the mature Weyburn oil field, where the captured CO_2 is injected into depleted formations to force out more oil.

The Boundary Dam CCS project is one of these legendary names that commentators pull out as proof of concept; these industrial personae come and do their dance. Drax. White Rose. Gorgon. Sleipner. Petro Nova. Kemper. If you read enough articles, these spirits of industry loom larger than life—they keep appearing, and journalists keep making site visits, trying to evoke them as interesting subjects. Some of them have already fallen; canceled ghosts. Yet it's hard to get a sense of their character; they are so easily abstracted as a collection of pipes. If you're lucky, they might make some kind of noise or smell, offer some sensory input to hold on to. Writing in the *New Yorker*, environmental journalist Elizabeth Kolbert manages the feat of evoking a direct air capture research facility: "In the workshop, an engineer was tinkering with what looked like the guts of a foldout couch. Where, in the living-room version, there would have been a mattress, in this one was an elaborate array of plastic ribbons. Embedded in each ribbon was a powder made from thousands upon thousands of tiny amber-colored beads." But when she comes to the Archer Daniels Midland BECCS plant in Decatur, Illinois, a geologist warns her: the facility is not sexy. She agrees: "It was, indeed, not sexy—just a bunch of pipes and valves sticking out of the dirt."[1]

The character of these projects emerges not from their pipes and valves, but from their settings. Boundary Dam and nearby Shand are set upon a scraped wasteland, with tangled little wild roses and thorns reclaiming the chalky ground. Mounds of white earth contour the surface, and blue water gathers in the hollows. This is the southeast corner of Saskatchewan, around Estevan, where the lignite coal seams of the Ravenscrag Formation have been mined since the 1850s. Lignite coal is softer, moister brown coal. It lies near the surface and is strip-mined. If you find a high point, a ridge or a picnic table outside of Boundary Dam, you'll be able to survey the coalfields stretched out below, the draglines resting. The infrastructure lives within a history, a sense of place.

Boundary Dam's claim to fame is groundbreaking, literally: it was the first power station to use CCS. On one hand, the technology's proof-of-concept promises updates; the renewal of aging infrastructure. On the other hand, the captured carbon dioxide finds its ultimate use in enhanced oil recovery, furthering extraction. It is a precarious moment for the facility. Although coal still powers 40 percent of Saskatchewan's electricity, coal with CCS can't compete with the flood of cheaper renewables and natural gas. While there's a feasibility study underway on whether to install CCS for Shand, at neighboring Boundary Dam, the decision was made to retire units 4 and 5 rather than retrofit them for further CCS: low gas prices make their operation a losing proposition. As of this writing, the federal government had agreed to keep them operating for a few more years without CCS. Kicking the can and the "stranded asset" label down the road until 2024 offers further job security to coal workers. If Saskatchewan makes the decision not to move forward with coal and CCS, 1,100 coal worker jobs could be at risk.[2] The whole thing is subsidized right now by a high carbon tax, which gets passed to consumers, who don't get to choose how their power is generated.

Indeed, despite being new and pioneering infrastructure, the status of "stranded" still looms over Boundary Dam—and indeed, many of the landscapes where these jobs exist already feel rather stranded in time. It's not as if stranding is an unfamiliar phenomenon here. There's a piece of a rusty, massive gear stuck by a roadside to commemorate Taylorton, a company mining town that's now overgrown with prairie—a casualty of strip mining technology that wiped out underground mines and the communities that grew from them. "Stranding" assets can seem politically inconceivable from the present's point of view, but it seems equally inevitable from the gaze of history, looking back at all the innovations and infrastructure which slipped into obsolescence because something better came along.

Is there anything likeable about projects like Boundary Dam, and what they promise? What, in fact, do they promise?

To many, a "good" future is synonymous with a "green" one. Forests, farms, the bioeconomy—intuitively, managing carbon goes hand in hand with tending the earth, massaging the carbon cycle by growing things. For life is already doing the work—it's a matter of encouraging it, restoring it, letting it flourish.

Through another lens, though, the lively biological realm is messy and unpredictable: it works in temporary time scales, and it's vulnerable to climate change itself. If the goal is permanent removal, far better would be to transform carbon in chemical and geological ways—the alchemy of the nonliving. Cultivation is generative. Burial, however, is pollution disposal, is safety, is sequestering something away where it can't hurt you anymore. One approach generates life; the other makes things inert. In this section, we will delve into the possibilities of geologic sequestration.

~

Carbon capture and storage is stuck. There is a major disconnect between what energy forecasters say is necessary (a lot of CCS), and what countries and industries find feasible (apparently, no CCS).

Carbon capture and storage is not a singular technology, but rather a practice that combines several verbs: *capture*, *transport*, *store*, *monitor*. The carbon dioxide is typically *captured* at the point where it's emitted. Generally, scrubbing the carbon out of gaseous waste streams is done chemically using compounds called amines, which bind carbon dioxide molecules. Then, the carbon is *transported* to where it will be stored, cooled into liquid form and moved by railway, ships, tanker trucks, and so forth, or conveyed as a gas via pipeline. This all takes a fair bit of energy. Basically, capture of climate-significant amounts of carbon dioxide entails an infrastructure on the same scale of today's oil industry—but to put the carbon back underground. The *storage* takes place in caverns underground, or in depleted oil wells. And then this carbon must be *monitored*, to make sure it is staying there. Eventually, it will turn into minerals, dissolve in water, or be trapped in rock.

The International Energy Agency (IEA) says that to meet a 2°C target, 3,500 CCS plants are needed by 2050. As of 2019, there are twenty-two large-scale CCS facilities expected to be operational by 2020—but this number has already shrunk, from seventy-seven planned as of 2010 down to forty-five in 2017, capturing a mere 80 million tons of CO_2 per year.[3] When it comes to what countries have laid out for how to cut their emissions under the Paris Agreement, only a dozen countries mention CCS (and none mention BECCS).[4] Yet the IEA's 2°C scenario has CCS delivering ninety-four gigatons of emissions reductions through 2050, half from the power sector and a third from industry, with BECCS delivering a total of fourteen gigatons of negative emissions during that time. "Without CCS, the transformation of the power sector will be at least USD 3.5 trillion more expensive," the agency cautions.[5] Do these reports get sent around just so the report writers will have done their jobs? Will CCS actually get going due to these new calls of urgency—or is the reputation of failure contagious?

Many green groups vigorously oppose CCS. According to Greenpeace, for example, CCS is a costly distraction that cannot save the climate. Enhanced oil recovery is "a euphemism for increasing oil extraction." Indeed, right now, the primary market for CCS is enhanced oil recovery: using CO_2 to force out oil that wouldn't otherwise be extracted. Greenpeace gives figures that without CO_2 injection, 65 percent of that oil would be left underground: "Under the auspices of helping the climate, carbon capture will be used to increase oil extraction by as much as 185 percent." They conclude, "It is clear that for the industry this is about extracting more oil—growing more as an industry—than they otherwise could."[6]

This is certainly true: for the fossil fuel industry, it is about extracting more oil. Yet there's also more to the story. This technology has received heavy public subsides; the public sector already has a strong role in it. The question is, what could CCS be if it were deployed not in service of the industry, but for us? Instead of reflexively dismissing it as a tool of the fossil fuel industry, we should work to understand its possibilities for helping clean up CO_2.

Perhaps industry's failure to make use of this technology could even be an opportunity to redirect it for more progressive ends.

To begin, let's look at a bit more of the history. CCS is another technology with roots in the 1970s. Cesare Marchetti, a modeler at the International Institute for Applied Systems Analysis in Austria (who also coined the term "geoengineering"), suggested storing carbon in the oceans, via a scheme he christened the "Gigamixer." His was one of the earlier writings proposing CCS to address the climate change problem.[7] However, throughout the 1980s, oil companies mainly explored carbon capture technology in the context of enhanced oil recovery, and pipelines for this end sprawled out from natural mines in Colorado to the oil fields of West Texas. At present, there are over 6,500 kilometers of CO_2 pipelines in the United States, chiefly for enhanced oil recovery.[8]

But in the 1990s, the climate community became interested in CCS. In conjunction with this, the United States and Saudi Arabia asked the Intergovernmental Panel on Climate Change to do a special report on the technology.[9] The resulting 2005 IPCC report put CCS firmly on the climate change agenda. In the lead-up to the 2009 UN Climate Change Conference in Copenhagen, \$30 billion in public funding announcements were made. But then, only \$2.8 billion was actually invested.[10] The mood soured; the hype fizzled. Cheap gas was once again king.

Many analysts think that the past few decades of research and practice have shown how to store and transport carbon safely, though others remain concerned. The CO_2 would be injected into deep geology—say, one kilometer below the surface—well below drinking water, into places like saline aquifers or empty oil and gas reservoirs. Storage capacity is not expected to be an issue. There are many places to put the CO_2—but this, along with storage integrity, is a topic of continuing research.[11] Notably, though, if highly concentrated CO_2 were to escape from a pipeline on land, this could lead to acid rain, acidification of water resources, and the asphyxiation of plants and animals; a marine leak could severely impact underwater life, as well. Concentrations of CO_2 above 2 percent have major impacts on human respiratory systems. At 7 to 10 percent, one falls unconscious and dies.[12] Obviously, pipelines need to be placed such that they avoid areas with important water resources, sensitive biomes, and seismic activity.[13] Induced seismicity, or the creation of earthquakes, is a concern that may arise with disposal of oil field brines into saline aquifers, which increases the pressure of fluids inside the rock. As stated in the National Academies report on negative emissions technologies, "The field does not fully understand induced seismicity from subsurface injection of CO_2."[14] This is an important research gap, and the report recommends allocating $50 million to study induced seismicity.

When it comes to injection of CO_2, global needs can conflict with local geographies. There's not always a "source and sink match," which is to say, the place where you'd want to build a carbon capture facility isn't always near the place where you can store the carbon. The cultural and social aspects of the terrain may constrain storage potential in many places. In India, one case study notes, the agricultural heartland of the Indo-Gangetic Plain has a large technical storage potential, but the arable land supports half a billion people. One of the suitable storage sites is close to the holy city of Varanasi, which would omit it from consideration.[15] Liability is also a huge issue: Who wants to insure this sort of risk?

If storing this waste underground for thousands of years doesn't sound great to you, you're not alone. Indeed, the forecast is still not bright for carbon capture and storage.

One of CCS's key failures has been its disastrous relationship with coal. Everyone thinks of it in terms of coal. "Coal for electricity," or imagining CCS's primary role as supporting continued use of coal for electricity generation, has been a hegemonic frame, as policy analyst Alfonso Martínez Arranz argues. This is especially true in Europe, he notes, where the dogma dangled the proposition of European coal's dominance over Russian gas, and promised a novel, emerging technology that Europe could offer the world.[16] This coal-for-electricity frame had its obvious appeal, given that fossil fuels powered modern civilization, that 80 percent of the world's energy supply still comes from fossil fuels, and that trillions of dollars have been sunk into this infrastructure.[17] Critics easily disparage this sort of CCS as a technofix that allows climate to be solved without any change in production or consumption. "The weakest point of the fossil fuel regime is presently its legitimacy," writes one group of critical scholars, observing that the addition of CCS promises to resolve this challenge.[18] However, they also suggest that the hype may have peaked: CCS has been contested by numerous NGOs, and it is still relatively unknown by the general public.

Even though it's tempting to see CCS as life support for fossil fuels, the truth is that coal plus CCS can't compete with cheap gas and renewables; it's a dead end in many regions of the world. Solar is a blazing star, while CCS for coal seems like a blast from the past—a holdover from the Bush I administration and its coal interests, belonging to a pre-financial-crisis era. CCS with natural gas looks far more economically viable. Many eyes are on a new demonstration scale power plant outside of Houston, Texas—Net Power's natural gas-fueled power plant. It employs a technology called the Allam power cycle that actually uses the heat of pressurized carbon dioxide to turn its turbines instead of steam—producing pipeline-ready carbon dioxide, thus bringing the cost of carbon capture way down. If it works at scale, it would be price-competitive with a regular gas-fired power plant, but without the emissions.

But let us think beyond these fossil fuels for a bit. Carbon capture and sequestration technology can, in fact, be used for all sorts of other things. For instance, it can be employed to curb emissions from heavy industry by attaching it to boilers and blast furnaces, steel mills, cement plants, and so on. I wouldn't be the first one to argue that CCS needs to be reconceptualized, and that advocates would have more success if they rebranded the technology to emphasize biomass, gas-fired power, or industrial uses.[19] We even see the IEA experimenting with these foci. The nonprofit Carbon180 talks about "renewable CCS," distinguished by two key features: (1) renewable systems that capture from an atmospheric CO_2 source (as in BECCS), or direct air capture from ambient air (as opposed to a fossil source); (2) long-term storage, rather than capture for short-lived fuel or chemical production. What you get, then, is not primarily an energy generation technology, but a pollution disposal technology.

There is an opportunity here to appropriate this group of techniques for redistributive ends. Morally, rich countries have an imperative to develop this technology, and use it, in order to reduce climate risk for everyone. This comes down to having an appetite for paying for and living with expensive infrastructure—and for making bright, clear distinctions regarding how and why it is built.

Reconceptualizing CCS is not going to be easy—especially given that the latest concept to gain traction is using CCS to produce low-carbon or carbon-negative oil. Essentially, enhanced oil recovery works by using water or carbon dioxide to flush additional oil from used-up wells. The industry has been doing this for decades, but now, there's a climate policy rationale for it. The key limitation to enhanced oil recovery, for the time being, is the low supply of carbon dioxide; most of the carbon dioxide the companies used is mined from caverns in Colorado and piped over to oil fields in the Permian Basin, in Texas. Installing CCS at ethanol plants instead, or building direct air capture facilities, could supply this demand for CO_2—and potentially more cheaply, especially if states like California certify the oil that is produced as lower carbon. Hence, a few industrial actors are promoting CO_2 EOR in residual oil zones as a "negative emissions" technology. In fact, direct air capture company Carbon Engineering announced a partnership with a subsidiary of oil company Occidental to build the world's largest direct air capture facility in Texas's Permian Basin, initially to capture 500 kilotons of CO_2 per year—to supply enhanced oil recovery operations, in a move touted as opening a pathway for carbon-negative fuels. Whether those carbon-negative fuels will ever be delivered is debatable, since they would only be carbon-negative under a particular life cycle accounting. It's easy to see how this maneuver could reassure nervous investors that there might be a continuing future for oil, and incentivize further oil production, while never delivering on the "carbon-negative" part of it.

Yet direct air capture technology need not be simply a tool of oil companies. The basic underlying technology could offer an opportunity to break the psychic chains between CCS and fossil fuels—depending upon who develops and controls it.

Mining the air

The idea behind direct air capture is simple. Instead of capturing carbon from a concentrated source like a power plant, or a cement factory, just capture it directly from the air. A handful of companies have working pilot projects that are doing this, using the captured carbon in greenhouses or to produce transportation fuels. Swiss company Climeworks has even partnered with a subsidiary of Coca-Cola to use its capture carbon in beverages. Yet direct air capture is expensive at present, in terms of both energy and money.

To understand the landscape for this emergent idea, I talked with Klaus Lackner, an engineering professor at Arizona State University and director the Center for Negative Carbon Emissions. He has been working for decades on direct air capture, approaching it as a waste management problem. Fossil fuel emissions are a cumulative problem, he underscores: the more you put out, the worse it gets. "See, we are actually all are incorrectly trained, and I actually don't know how that happened. It is in itself an interesting question," he muses, explaining that sulfur pollution and acid rain in the 1970s and 1980s effectively trained us that pollution has a lifespan in the atmosphere, and then goes away. CO_2, however, truly accumulates. "And it actually does change the way you think about the problem. You see immediately, if I pretend garbage is CO_2, right?" You wouldn't allow people to throw waste in the street, he says, and you wouldn't accept it if someone claimed they were just dumping 10 percent less garbage than last year. We can't afford to dump CO_2 in the atmosphere, he says, yet we have been. "But now the litter is so, so big that even if we stopped emitting tomorrow, we have to come back down. So we need carbon removal technologies."

What exactly is he building? Picture a machine the size of a shipping container—each one removes a ton of CO_2 per day, and you could deploy four of these on a square kilometer. Sequestering 4 million tons per year on that square kilometer would be like erasing the emissions of half of a big coal plant. These machines would be mass produced, and situated in places where you'd want to collect and store CO_2. They would provide a service—the cleanup of pollution that was ignored in the past—much as the engineers who maintain an urban sewer system provide a valuable service.

One engineering challenge is that carbon is dilute: 400 (plus) parts per million. Lackner suggests thinking of windmills as miners that pull kinetic energy out of the air. With direct air capture, pulling the CO_2 out of the air is like mining it. If it's mined for disposal, perhaps all he gets is a tipping fee, a fee for the waste disposal—say thirty dollars a ton. "So now I can say, 'Okay. I have a cubic kilometer of air.' Which sounds huge, but it turns out a big windmill sees this in an afternoon. You can ask, 'How much kinetic energy is there?' Turns out, it's $300 worth. If you ask how much CO_2 is there, turns out it's $21,000 worth." Even though CO_2 is dilute, it's relative— the CO_2, by this measure, is seventy times as concentrated as wind energy.

And, he adds, the geographic footprint would be smaller than that of wind energy, especially if you wanted to fill a single drill-hole with carbon. "You could think of these things sitting around in corners of a field. You could see them in corners of a wind farm. And they are actually smaller in the wind farm. Or solar panels." They could absorb excess capacity when supply of renewables is higher than demand. Imagine looking at it from an airplane: most of what would be visible is simply solar panels. "You see these solar panels there; there's this one curious thing in the corner, which is there to collect the CO_2." The footprint might be physically smaller than the renewable energy sector, but the revenue might be comparable.

In this scenario, scaling up direct air capture sounds much like the scale-up of renewables: rural and remote communities are asked to bear the burden, with some possible benefits. I wonder if there's any way to make the infrastructure more lively or beautiful—the devices have been referred to in the press as "artificial trees." Is it actually possible to actually design air capture devices that look like trees, I ask? He says that it is possible, but his team is probably not the right group to do so—and gives an explanation involving a membrane and the difficulty of making a flat sheet of his material (because it expands when it gets wet). The main design consideration, though, is to site direct air capture where you want to store the CO_2.

"The architecture group here wants to do fancy things on the outside of houses and buildings. And they say, 'Well the wind blows by there, why don't we just collect CO_2 while we go? And being in Phoenix we can use them as shade structures at the same time.' Then, well, now you have ten tons of CO_2 at the end of the day out of your building; what are you going to do with that? Right? I need to be somewhere where I can make use of that CO_2. So I think a lot of them will be like your mirrors in the desert, like your windmill farms in Kansas. And they will just be out of sight for most of the population. I think that's actually very likely. And not only because we want to hide them, but also because most of the places where you really could do something with that CO_2 are not under cities. For one thing, people don't want to live right on top of the CO_2 you just put away. And this gives you the flexibility, right?" Lackner notes that this would inherently be a very big industry. If you could do it for thirty dollars a ton, which he names as an ambitious target for the long term, you'd need a trillion-dollar-per-year industry to pull as much CO_2 out as we are currently emitting, meaning that it would take us forty years to pull back one hundred parts per million. But, he adds, it would also be an industry with a known sunset, because we wouldn't want to take out, for example, 300 parts per million. "And so, you know at some point that task ends."

Critics are quick to point out that the thirty-dollar-a-ton target is very aspirational. A 2011 report by the American Physical Society on direct air capture, one that still haunts discussions of the technology, projected it would cost $600 a ton.[20] (In fact, this is what it costs a company called Climeworks to do direct air capture at its small demonstration facility in Switzerland, where the CO_2 is used in greenhouses for growing things like cucumbers.) Lackner critiques the methodology in this report, though, because it is based on a conceptual device made entirely of off-the-shelf things that we already know how to do, with no attempt to make the process streamlined and efficient. "Once you get on an experience curve and start learning," he says, "I would have taken those messages as saying, 'Oh! This is actually a remarkably good start for something!' We aren't done yet. Right?" Lackner thinks that bringing the price down by a factor of ten might not be that hard, ticking off examples of technologies that get cheaper over time, especially when using mass production: photovoltaics today are one hundred times cheaper than in the 1960s, and wind energy is about fifty times cheaper.

As of this writing, I think it's fair to say that we don't actually know how much it will cost to capture carbon from the air. The system designed by Carbon Engineering, a firm based in British Columbia, reported costs of between \$94 and \$232 per ton of CO_2.[21] If costs could be brought down to one hundred dollars per ton, removing five gigatons per year would run about \$500 billion, or 0.6 percent of global GDP. (For comparison, the bill for damages from the US hurricane season in 2017 was about half that.) The primary limitation of direct air capture technology is financial—a reflection of people's willingness to pay to dispose of this waste properly. It's a social value. "Look at what happened in recycling," Lackner suggests. "How did people decide that you actually recycle, and not turn around the corner and dump it into the landfill?" People who you can't convince of the reality of climate change, Lackner observes, can still be convinced to clean up after themselves. He recalls an open house where his team set up a table with Ziploc bags filled with sand: a half pound, a one pound, and a two pounder. They'd ask people how many miles their car gets to the gallon; if it was thirty, for instance, they were handed a one-pound bag of sand, which weighs the same as how much that vehicle emits. "And most people were shocked how much stuff comes out of their car. Because it's colorless and odorless and not visible. So they don't appreciate how much it is."

When we think from this waste management frame, it makes it easy to connect the nascent direct air capture industry to environmental justice concerns. Lackner sees all kinds of regulatory questions that will need to be answered as the industry matures: the definition of a true removal, how to know it was done, how to know it is permanent, and what industry standards should govern it, to name a few. There are safety issues, too: the process needs to be certified as safe and harmless. "This then ties to policy, ties to regulatory frameworks, and ultimately ties to the law. If I'm downwind from a field, can I get sued by a farmer who says his corn didn't grow as well as before, because I took CO_2 away? Nobody has thought about that yet."

The environmental justice questions aren't just a matter of avoiding injustice. Could a direct air capture industry could further climate justice? If so, what would it take to accomplish this? I ask Lackner about how his technology could address things like historical responsibility.

"Yeah, you could imagine if you were so inclined ... We get started, and because we're not very good at it, this year we clean up 1800 to 1804, next year we are cleaning up 1804 through 1808. And we keep going along like that, when we hit the 1950s we really have to be big, right? And so, I see the argument India has made, that we should not stop their development." But as Lackner, notes, we have never been good at reparations. "Think of slavery, think of colonialism. We even refuse to apologize, because it would set precedents. So I have a hard time seeing that the industrialized countries suddenly see the light and turn around and say, 'Okay, it was all our fault, we will take care of it' ... So I think the outcome is much more likely that we handle it on a per capita basis today." There's an argument to be made, though, that carbon removal could be reparative. "If every person on the planet has a carbon budget which includes his ancestors' carbon budget, then we in the US and in Europe have basically blown our budget a long time ago, I mean. Carbon removal technologies, air capture in particular, would actually allow you to pay it back. So you could negotiate and say, 'We know that we had a history of it, and we will level the playing field.' You could, in principle, say that. Am I an optimist that we will say that? No."

But Lackner is more optimistic that we can learn to see this as a waste management problem. "We have taught people about recycling. We have taught people about renewable energy. And there we were actually extremely successful." He suggests that if there were a certification system that was trustworthy, people might be convinced to pay a little extra for carbon removal. If you could convince Amazon, for example, to send its customers a pop-up at the moment of payment that says: "'Your CO_2 bill would be fifty cents.

Are you willing to pay that?' And I'm sure I would say yes, if I trust the system is actually doing it." Just like recycling, Lackner thinks, it will start with a few standalone adopters and volunteers who feel strongly about it. "At some point, politicians say: 'And now it's regulated. You don't have a choice in the matter anymore. If you want to buy a gallon of gasoline, twenty pounds of CO_2 have to be put away.' And then you could also have volunteers who say, 'I'll take my grandfather's CO_2 back.'"

Lackner is clear to say that at the end of the day, he can't guarantee that this technology will work. "I think it would work, but until it really works, I can't guarantee it. I am the first person to agree that we shouldn't take it as a fact that we can do it, and therefore that we have nothing to worry about. But I can tell you that if it fails to mature in our lives—because we have not worked on that, or maybe because we couldn't make it work—then there will be a high price to pay for that. Because we will overshoot, and if we can't come back down, we will suffer all the consequences. If we can come back down, the consequences will be less."

Taking CCS to the next level

What are the costs of failing to understand CCS through the right frame? Without active engagement from climate justice advocates, CCS probably won't go in the right direction, and an opportunity for climate repair will be missed.

Direct air capture can be applied to carbon removal in two fundamental ways—what I'm calling the two levels of carbon removal, where the first level involves million-ton scales and the second level involves billion-ton, climate-significant scales. On the first level, direct air capture could be used for so-called "carbon capture and utilization," or CCU: niche applications in the "carbon to value" economy, or carbontech sector. A lot of this is basically what I'd call concept art, where the value is getting people to think differently about carbon: *Did you know you can make shoes out of carbon dioxide?* Indeed, materials scientists have discovered all kinds of things you can make with captured carbon dioxide, from makeup to carbon fiber. Some people working in climate policy see CCU as a step stool from level one to level two. Much like with enhanced oil recovery, the idea is that it provides an initial market to develop the technologies and bring their cost down. One consortium trying to bolster the carbontech sector, the Global CO_2 Initiative, claims that the CCU sector could reduce emissions 10 percent by 2030, identifying four major markets: building materials, chemical intermediates (e.g., methanol, formic acid, syngas), fuels (methane), and polymers.[22] This might actually be true, if you were to count the emissions displaced by the decarbonization of concrete manufacturing; cement production, it turns out, is responsible for some 8 percent of global emissions.[23] Indeed if cement were a country, only China and the United States would emit more. so decarbonizing concrete is promising. And air-to-fuels, as mentioned earlier, hold huge potential as a use product for the captured carbon. But again, avoided emissions are not the same as negative emissions or carbon storage.

However, this scale of projects is vastly different than Level 2 carbon removals: removals that would affect the climate and actually draw down greenhouse gas concentrations in a noticeable way. The key limitation of turning carbon into valuable products is the staggering scale of current emissions—so much carbon is emitted that we'd only be able to use a tiny fraction of it. A 2017 study in *Nature Climate Change* examines the potential contribution of CCU to climate goals. The authors find that "CO_2 utilization is highly unlikely to ever be a realistic alternative to long-term, secure, geological sequestration," and that it could comprise only about 1 percent of the mitigation challenge.[24] Another issue is that a lot of the products release the carbon again at the end of their lifespan. If carbon is turned into urea and used as a fertilizer, or turned to methanol for fuel, it's only utilized for some months. The authors strongly emphasize the danger of reinforcing the narrative that CO_2 utilization is key to CCS. "If this narrative continues," they warn, "it introduces the very real risk that emission mitigation targets will not be met and that CCS through geological storage will not be deployed in any meaningful way."[25]

To understand the perspective of an entrepreneur in the "new carbon economy," I spoke with Tito Jankowski. He's a founder of the sustainable tech consultancy Impossible Labs, which also hosts airminers.org, an index of companies mining carbon from the air. Jankowski emphasizes the importance of tangible products for generating demand for air capture. "How do we put this stuff in people's hands?" he asks. People find it difficult, he notes, to believe in climate solutions that look like a machine out in the middle of nowhere that nobody sees. "How are we really supposed to enthusiastically buy this stuff if we call up the company and they say, 'Oh, okay, yeah. You need $400,000 to buy one of our machines.' How are 7 billion people supposed to get behind that, and say hell yeah?"

To make it tangible, Impossible Labs had limestone containing carbon captured by the Carbon Engineering, the direct air capture firm, shipped from British Columbia to Oakland. It then molded the limestone with the captured carbon into a consumer product: a planter. Cofounder Matthew Eshed recalls the genesis: "Yeah, it was something we just kind of came up with … you put a plant in the planter, and then a plant sequesters carbon as it breathes and grows, and photosynthesis happens. So it's kind of a nice full story contained in this one thing," Eshed describes. The planters were quite popular, though; they all sold out within an hour. Jankowski asked people: Why did you pay one hundred dollars for this cup? "And people said the craziest stuff. Like, 'This is a symbol of the future.' Somebody said, 'This is what I want to see more of in the future, and buying this thing is the best way I can think of to send a signal to the universe that I want more things like this.'"

Part of the effect of seeing the captured carbon, though, is reckoning with the scale of the challenge. Eshed describes learning through the process of making this artifact. "What happened for me was, I saw the scale that atmospheric carbon removal needs to achieve anything that's measurable, and that's what really kind of caused me to start going in a different direction," Eshed says, commenting that he's now focusing his professional life on local climate resilience. "What we heard from people is that it represents a conversation. It's a place for a conversation to start, and whatever industry you're in, you can look at this thing that we've made. We say, 'This product has 116 grams of carbon dioxide from the air inside of it,' and it's like, 'Well, that's not really anything, let's do better. Can you do better? How can you?'"

Taking direct air capture to the second level, where it works as a pollution remediation mechanism that makes a difference in the climate, is going to require enormous, transformative political action. Who should pay for it? Using a logic of pollution control, it's the fossil fuel companies that should have to. This might mean not just getting rid of these corporations, as we might like to, but transforming them into companies that deliver a carbon removal service. We can anticipate having to subsidize their transition. There will be a lot of struggles to engage in here. Yet it wouldn't necessarily have to be incumbent fossil majors that perform the removal service—it doesn't have to be a vertically integrated scheme. Geologist Stuart Haszeldine suggests that extraction should be linked to storage, using obligation certificates. Each ton of fossil or biocarbon production would be allocated a certificate by its originating government, and the storage obligation certificate would create new CO_2 storage businesses. Hence, the extractors or importers wouldn't need to store their own precise tonnage, but they could pay a services company to store the equivalent—thus, a lever to enforce storage that doesn't rely only on taxation.[26] Direct air capture companies could be the ones fulfilling these storage services.

Green groups could push direct air capture to the next level—if they decide to engage with it. Is there room to demand cleanup action, a remedial or restorative politics that uses direct air capture or other forms of CCS for climate justice? Without the involvement of civil society, the prognosis looks grim. Anthropocene theorist Kathryn Yusoff asserts that "it is through the violent infrastructures of geology that new forms of politics are emerging, such as those at Standing Rock around the Dakota Access Pipeline that insist of a different vision of temporal affiliation and material filiation."[27] What, then, would be the infrastructural politics of CCS, besides resistance? Yusoff refers to cultural theorist Lauren Berlant's work on infrastructures as convergences of force, structures of feeling; pointing to the affective dimensions of subterranean infrastructures as well as their instability. However, I think that reformist green capitalism or energy-forecast wishes are not made of strong enough feeling to force a vast, deep carbon clean-up infrastructure into being. This genesis instead needs to be driven by the desires of people—people who at this point have no idea that CCS exists, or who have learned antipathy toward it based on its coal tones.

Living with gas and liquid-filled caverns of carbon isn't the only option for geologically storing carbon dioxide, however. It's also possible to store it in a solid state, by turning it into rock, which is what we'll look at next.

Sketch: Pecan Tree

He was sitting under the old pecan tree, when a truck pulled up in the lane. Utility people. He leaned on the armrest of his rusty lawn chair. "Wasn't expecting you folks," he called.

A man and a woman briskly got out, wearing spring-green uniforms and tablets at their belts. "Your account wasn't taking messages, Jack. Full up."

"Well, I haven't got time for messages." He wiped his brow, stood up. "Just taking a rest under this pecan tree a moment. I was going to go check on unit eight this afternoon."

They looked at each other. "Jack, that's what we've come here to say," the man said, taking off his cap. "We're decommissioning your ten units. This formation is pretty well full up. There was some seismic activity nearby a few weeks ago. The modelers don't want to overstress the rock."

"Haven't heard anything about this."

"You haven't been checking your messages, Jack. We even sent a drone out with a message last week."

"That metal bird? Shot that thing down once it crossed my property line." The woman shrugged.

Jack crossed his arms. "They told me those would be operating my whole lifetime. I grew up with those wells, you know. I was there when they got put in."

"You must have been a kid."

"That's right. Came home from school, and my parents were serving iced tea to the utility men on the porch. I remember their excitement. Said we wouldn't have to worry about money anymore. I could have gone off to study anywhere, after that."

"You know, Jack, even though your repair contract ends when we decommission the units, you'll still be getting storage rent. It's not as much, of course."

They stood there. The birds in the tree were crowing up a storm. They did that when outsiders came around. "We'll just be heading out to the units to package them. There'll be a rig around to ship them off tomorrow."

"Well, I should be the one to turn them off, at least. You'll need help with the cover on five. It has a special trick to it. No one knows how to remove it." He ambled over to the shed to find his toolbox.

It was a cloudy day. "I can't imagine living out here among all these wells," said the woman.

"Why not?" replied her companion. "There used to be an orange-brown haze across this whole panhandle, my mom said. She always blamed my grandma's asthma on it. Now the air is clear."

"It's clear, but there are still these pipes all over the place. And dried-up roads crisscrossing everything."

"Best hunting, around here, 'cause there's less people. They even opened a savanna across the state line. With camels. Oklahoma camels. You can hunt wildebeest."

"That's horrible." She wrinkled up her face. "Why not bison? Those would be native, at least."

"I dunno. I guess that was too close to home. I mean, just half the savanna is open for hunting. The other half is preserved, you know. That's the deal they struck. To get people to sign over their land. Anyway, what else are you going to do with all these old oil fields?"

"Leave them. Let nature do its thing."

"Yeah, but it's up to the people who live here. If it was up to you, we'd all be living in cities zipping around in tiny scooters. Me, I took this job because it was one of the only chances I had to drive a truck with my own two hands."

Jack came meandering back, toolbox in hand. "Ready to go," he announced.

The first five units were on the other side of the ridge. They squeezed into the truck, rumbling over a dry creek. Yellow cottonwoods brushed the brown grasses. Jack spied a few weathered boards still tacked to a trunk from the treehouse he'd built for his daughter; echoes of her hair in the wind.

The first unit was home to a nest of swallows. The weathered metal container sat next to a drill pad, a footnote to the massive wind turbines lining the ridge. "These swallows have been here about five years," Jack said. "If I'd known you were coming, I would have come up here and built them a birdhouse." Jack keyed in the code and unscrewed the casing. "No mouse droppings. These sonic repellents I put in are doing the trick."

The man inspected the control panel. "This is a real vintage model."

"Yup. One of the very first in the world, out here in the Permian. Gave it a few updates over the years."

"Well, I guess we'll shut it off."

Jack stood there for a moment. "That's all, then." He began to flip the switches. The fan on the unit slowly wound down. The techs were absorbed in their tablets. The blurred blades quieted and then stopped.

"One down," the woman said.

They drove the loop road, a track Jack had pressed into the ground through his rounds over the years. The last one, ten, was his favorite. He used to bring his oldest son up here to watch birds. Jack unscrewed the panel, flipped the switches, and watched the blades still. He looked down at his crusty boots, the leather starting to crack.

"I got the final number," the man said, looking up from his tablet. "These units captured 406,781,200 tons of carbon over their operational cycle. That's almost all of Texas's emissions from the year 1980."

"You should be proud, Jack," added the woman. "That's good work, for a lifetime. A whole year's worth, and not one of those early years, either."

"Well, I figure so," Jack said. "I guess that's that."

"What are you going to do with all your free time, Jack?" the man asked.

He was silent for a moment. "I guess I'll work on some hobbies," he said. "Taxidermy. Maybe do more fishing."

"Sounds good," the man said, nodding. "Sounds like a plan."

Jack shook their hands solemnly before climbing out of the pickup. Then he stood under the pecan tree and watched the two drive off, small puffs of dust dissipating slowly into the cooling air.

5

Weathering

Los Angeles, January, 30°C / 86°F

The eventual fate of carbon is stone. In hundreds of thousands of years, the atmosphere would naturally return to conditions like our ancestors knew: minerals will take up excess carbon dioxide. Very slowly. Here's how the basic process works in nature. In the atmosphere, carbon dioxide reacts with water to form carbonic acid. Slightly acidic rain will dissolve rocks on earth's surface, forming inorganic carbonates which eventually wash into the oceans. There, shell-building creatures and plankton turn the calcium ions into calcium carbonate, and over time, the built-up layers of shell and sediment turn into solid limestone. This gradual weathering process naturally sequesters about one gigaton of CO_2 per year in the resulting solid rocks. Though humans are emitting forty to fifty gigatons per year, one gigaton of capture is still a significant contribution by these little-noticed rocks.

This begs the question: Can this natural process of rock weathering be sped up?

"Enhanced," "accelerated," or "high-speed" weathering all refer to hastening natural reactions that break down rocks and eventually create carbonates. (Make Nature Work Faster could be a slogan not just for engineering super-algae, but for rocks as well: What can't be sped up?). For more information on how one would go about turning carbon into minerals, I paid a visit to Joshua West, a geologist who looked into enhanced weathering for geoengineering while a research associate at Oxford, and who is now a professor at the University of Southern California.

Los Angeles was in the throes of a January heat wave. The sun was blazing, but people breezed by on their skateboards, scooters, and bicycles under the azure sky. The geology hall is a Romanesque stone building. Its arcade features a "sacred gifts of nature" shelf, inviting passersby to "leave something and receive something." The sacred gifts—shells, fragments of driftwood, carefully arrayed—rest serenely beneath a mural of a deer and four contemplative humans, staring wondrously at an orb filled with squiggly microorganisms. Passing through the sunny courtyard, quiet save for birdsong, you could forget that the neoliberal, entrepreneurial university even existed, and imagine you'd stumbled through the heat into a temple of natural worship. The halls were decked with artfully displayed slabs of 2-billion-year-old rock surfaces like pink "Vyara migmatite," swirled from potassium feldspar. "Geochemists Only" read the sign on the door of the lab. West answers my knock. He's an affable and extremely tall man with a bright-violet shirt, jeans, and brown leather shoes which, from the looks of them, have certainly seen many interesting rocks. He invites me to sit down on a couch, across from a Phantom remote-operated drone and rolled-up posters detailing discoveries in rock weathering.

West has been intrigued by earth's long-term carbon cycle since the beginning of his career: "What is the mechanism that kept climate stable, and what have been the times when it's been deviated from that stability?" He's spent his career working on things like the massive volcanism that led to extinctions—the big questions geologists love to ponder. Today, West researches a whole host of questions unrelated to geoengineering. But given his long expertise in weathering, I asked him to go over what exactly it is.

West patiently explains that in nature, there are rocks called silicate minerals, in which calcium and magnesium are bonded with silicon and other elements. Rainwater falls on them and reacts with them, slowly breaking them down and releasing the calcium and magnesium. These calcium and magnesium ions eventually flow down into the oceans. "If there are ways that we can make the breakdown of those calcium and magnesium silicates go faster, then we can produce more calcium and magnesium ... and in an industrial context or something like that, we can make chalk, basically. We can make all the carbonate minerals that would be made naturally over 100,000 years, but rather than waiting 100,000 years, we do it fast. And if we can do it fast enough, then we can meaningfully reduce the amount of CO_2 in the atmosphere."

How fast is "fast," I wonder? Faster than 100,000 years could still be pretty slow ... are we talking about ten years, one hundred years, or 1,000 years? In a sense, West explains, it depends on how much money and energy we are willing to put into this. It also depends on the technology used.

Turning gas into stone

One way to speed up carbon mineralization is to inject concentrated CO_2 with water into rock at high temperatures (in mineral weathering parlance, an "in situ" method, as it takes place within the rock, underground). This technique was used at Iceland's geothermal Hellisheiði Power Station, at a project called CarbFix, where CO_2 that bubbles up with the magma that fuels the geothermal plant was mixed with water and hydrogen sulfide—carbonating it, essentially into "seltzer." Then, the mixture was injected into basalt rocks 400 to 800 meters below ground. Basalt is the most common volcanic rock—it underlies the oceans—and when filled with this soda water, its pores fill up with limestone. The results? Ninety-five percent of the injected CO_2 had turned into mineral within two years.[1]

Turning carbon dioxide into rock underground seems much more attractive than storing it as a fluid, from an intuitive standpoint. It's permanent, for one. West says, "I do think that there's a sense of security in turning CO_2 into a mineral form, one that isn't gained by trying to put it somewhere where it's still a fluid, 'cause fluids move. Minerals move, but they're solid." The CarbFix project has even installed an air capture device, moving it toward being a true negative emissions technology.

Inevitably, there are questions about scalability, especially around the energy and financial costs. This particular project in Iceland is also very water intensive, requiring about twenty-five tons of water for every ton of CO_2. This means it might not be a good option in semiarid regions. One option could be to do it offshore—where it would have a ready supply of water, and would not conflict with land uses. But working in offshore environments would also increase the expense. The most suitable areas are 200 to 400 kilometers from land, necessitating pipelines, and their depth is around 2,700 meters, meaning that even a demonstration-scale project would be quite costly.[2] Nevertheless, this is an exciting new area of research.

Rocks for crops

The other way to enhance weathering is using so-called "ex situ" methods, which involve grinding up rocks to a fine powder and turning them into carbonate minerals. One way to do this would be in industrial facilities, using heat or acid and running CO_2 through the rock powder. "That, we could do very fast. I mean, that, you could do in as long as it takes you to build the facility and dig the rocks up."

The other ex situ approach involves excavating rocks, grinding them up, and spreading the rock powder on fields. This is colloquially known as "rocks for crops." The rocks-for-crops idea enhances weathering by increasing the surface area of the material via the crushing and grinding. The biological activity in soils also can speed up weathering. (Again, we witness the wonders of soil microbes!) Ground-up rocks dissolve one to five times faster when plants are around.[3] West remarks, "It's actually a pretty interesting idea because it has the potential benefit of providing nutrients to crops, too, so you can get a double whammy, especially in developing countries where fertilizers are a little more scarce or more expensive relative to the value of the crops." Right now, a research team from the Leverhulme Centre for Climate Change Mitigation in the UK is conducting mesoscale field trials in Illinois, Australia, and

Malayasian Borneo to figure out if this is something that works on a five-to-ten-year timescale.[4]

There are some respects in which rocks for crops looks like a win-win solution. It is a fairly low-tech approach, and it could be done right now with existing technology. But it would also be a massive mining and transport operation, with its own particular geography. "We have to dig up rock, and so that's a big mining operation; and another aspect here is that some rocks are more reactive than others, so you can't just dig up any old rock and expect it to work very well," West explains. You would have to go where the rocks are—places with minerals like olivine, one of the faster-reacting silicates. Certain tropical areas, with warmth and moisture, look the most favorable. In short, rocks for crops is attractive in that it lets nature do the work, that it could have co-benefits for farmers in terms of fertilization, and that you don't have to concentrate CO_2.

Of ultramafic silicate rock flour, and mountains

Rocks for crops faces questions of scalability, as all these carbon removal techniques do. The scientific literature is full of calculated potentials: for instance, that dusting two-thirds of productive crop-land with basalt could extract 0.50 to 4 gigatons CO_2 equivalent per year by 2100.[5] The aforementioned bestselling compendium *Drawdown*, on the other hand, suggests that if olivine was applied to one-third of tropical land, it could lower atmospheric carbon dioxide by 30 to 300 parts per million by 2100.[6] But *Drawdown* also notes that some scientists claim the rates in nature are actually ten to twenty times higher than those in the lab. I ask West: What is going on with these estimates?

"There's this fundamental problem in what's called weathering kinetics, which is that there's an enormous discrepancy between lab and field rates," West explains. Laboratory experiments can't capture all the effects of biology on mineral dissolution. "If you add even microbes—forget plants and all the assets they excrete—just the little molecules that microbes excrete are enough to dramatically, potentially ten, twenty times, increase the rate that minerals dissolve … So, we know that. We don't know exactly why. What is it that microbes do that make minerals dissolve faster?" Field trials will help reconcile these discrepancies between lab and field rates of weathering, but they're not an easy thing to get funded, because they're considered "applied science" for this particular goal. That is, you'd have to be studying rock weathering specifically for carbon sequestration purposes—something that is beyond basic science. Most funding agencies aren't quite to that point, save in the UK, where there is some dedicated funding for this research. However, we also need fundamental basic research into areas like reaction kinetics, which will be crucial to understanding the results from field trials.

Is it possible, I ask, that field trials will show that enhanced weathering is a way bigger solution than you read about in, say, assessment reports on carbon removal?

"I think there is certainly the possibility that field trials will show that it's more effective than we've guessed, right? But there's the possibility that it goes the other way around."

West then points out a few more complications with enhanced weathering. Olivine—a rock that is more effective than basalt at sequestering carbon—contains trace amounts of nickel and chromium that are toxic, and so spreading it on the food supply, or on tropical forests, might cause it to accumulate in the food chain. This is one reason why researchers are now looking more at basalt, even though it's less effective for CO_2 capture—about 0.3 tons CO_2 per ton of basalt, versus 0.8 for olivine.[7] Another consideration is that the calcium and magnesium get released from the rock, but then bond to clay minerals—"secondary minerals"—instead of ending up bonding to CO_2 to make carbonate and increasing the amount of carbon stored. Finally, the literature points out another potential issue, which is how the application of rocks would impact biodiversity in the surrounding ecosystems—biomes that may already be adapted to nutrient-poor and acidic mature soils.[8]

Now I'm ready to ask West my dumbest question: If we start drawing massive amounts of CO_2 into rock, are we going to end up with too many rocks everywhere?

He laughs. "Oh, no, you can't do that."

"Okay. It's the type of thing that people might wonder about."

"Okay. I say that glibly, that you can't do that." But West recalls a geochemist group expedition to Oman, where there are mountains where the ancient seafloor has been uplifted. "And that ancient seafloor—there's huge mountains of it that had been converted into carbonate, and we tried to do some back-of-the-envelope calculations on what's the size of rock that would be needed to take ten gigatons of CO_2 or whatever, some reasonable amount of carbon to offset emissions. It's enormous. It's a mountain's worth, right?"

"So, each year," he continues, "you have to produce a mountain's worth of carbonate somehow, and that could be in the oceans. The oceans are big, right? So, several mountains in the ocean ... it sort of disappears under the scale of the ocean; but if you're thinking about doing this industrially, it's actually a lot of carbonate that you have to make." West explains, again, that you have to start with some matter—you have to dig up rock, because you're taking one rock and converting it to another. "There's a conservation of mass. You can't just make new rock, so we're not gonna bury ourselves under rock by doing this. But that said, the calculation that one year's worth of CO_2 emissions amounts to a couple of big mountains' worth of material, and I'm not talking, like, just piles. I'm talking a big mountain. I don't remember what the numbers were, but I remember many kilometers. That, I think, serves to illustrate, at least in my mind, what a huge challenge it would be."

"The fossil fuels industry has an enormous footprint, and we don't think about it on a daily basis. But you just go to a refinery in LA and it's huge—and they're *tiny* compared to refineries in the Middle East and so on, and then you add them, and it's just this enormous industry. Effectively, if we want to offset that in an industrial way, we have to have an industry that is of equivalent proportion, and I think that's sort of lost." The engineering side of it is pretty eye-opening to think about, West says. "I say that we could do this in five years if we wanted to. Well, sure, but are we really gonna build an industry the size of the fossil fuel industry in five years?" If there's an abrupt change in climate, we might be stimulated to do that, he speculates, but gradual climate change might be hard-pressed to kickstart that kind of industry. At any rate, the scale of mineral carbonization is enormous —"but we won't bury ourselves. Does that make sense?"

I'm glad we won't bury ourselves under mountains of waste carbon, but my questions about scalability are lingering. One aspect I've been wondering about is the competition between technologies: If we're going to build out a fossil-fuel-sized industrial infrastructure, which is what we talked about earlier with BECCS and direct air capture, would we choose to build it out for direct air capture or BECCS, rather than for enhanced weathering? Interestingly, there are some ways that enhanced weathering is complementary to other carbon removal practices. For one, biochar is proposed as a carbon sequestration measure that would have fertilization co-benefits in the tropics; rock dust could be complimentary to it, because biochar alone doesn't provide enough nutrients. Second, enhanced weathering could be used on bioenergy croplands, if those were expanded for BECCS. And third, enhanced weathering might be beneficial for large-scale tropical reforestation programs.

So there are some overlapping spaces where enhanced weathering could complement other drawdown approaches. When geographical factors are included, however, they shrink the potential. One study points out that 80 percent of agricultural commodities are consumed locally, and that areas with limited exports wouldn't have the transport infrastructure to import basalt for spreading on their fields. Some basalt is held in outcrops without arable land (e.g., Siberia, drylands of Ethiopia), and there's a carbon cost to moving the rock around. The authors argue that the initial deployment of weathering on croplands would take place in areas with good road access, heavy machinery, and basalt nearby, like North America or the UK.[9] And it's not just road and transport infrastructure that's needed— labor is also required. Crop, tree, and rubber plantations managed by large-scale agribusinesses that already apply crushed limestone and fertilizer would likely be the first adopters. Small-scale farmers may not have the resources. But the counterargument is that as agriculture develops toward large-tract mechanized farming rather than smaller-scale shifting cultivation, and as roads are developed to reduce yield gaps and bring new cash crops to market, agricultural systems will transform into forms more compatible with enhanced weathering.[10] To what extent is this vision incompatible with smallholder, agrocological, polycultural food systems? Even if mechanical methods are available, will women still be bearing bags of fertilizer, as they are in Indonesian plantations?

I once attended a presentation by some modelers who showed a world map of the areas that would be promising for implementation of rocks-for-crops. The best spot, it seemed, centered in the Democratic Republic of the Congo. I raised my hand and asked how you might get people in the DRC to use mineral weathering on their fields. Their answer: "We'll just pay them." This might seem like a reasonable approach to one who is unfamiliar with the operational realities of the DRC. But I will leave this here.

As with many climate engineering ideas, one could see this concept as either promising or terrible, depending upon how it is implemented. Mining is often deeply disastrous to both ecosystems and communities. Spinning up a new mining industry could alter the social fabric for the worse—but on the other hand, it could be a source of employment for people with expertise in other types of mines we might want to be phasing out. We could imagine silicate mines being hazardous, dust-inducing places if they were not built out well. One paper flatly states, "Especially in areas where agriculture is not managed by agribusiness, this would require a pan-tropical investment in education, safety equipment and protocols."[11] Like with most other geoengineering ideas, so much depends upon how it is done, and by whom.

What on earth would motivate the creation of a new industry for removing carbon? It's the same question we considered for CCS in the previous chapter—but its potential proponents are quite different. When it comes to enhanced weathering, some of the research is done by geologists such as Dr. West, who are not advocates of going out and doing this recklessly or at large scale, but rather are interested in the science, and what we can learn about our earth through exploring it. Then there are modelers, who come up with amazing maps and calculations about the potentials and costs, but are perhaps overly driven by the need to produce papers. Enhanced weathering is quite like BECCS in this regard: it could be an artifact from a community of modelers who are required to create interesting and important projections to keep their jobs. But enhanced weathering lacks obvious champions to operationalize it. One interesting move is a research project by diamond multinational De Beers to implement accelerated weathering to store carbon in kimberlite rock at its mines, given that it has all these mine tailings lying around. A scientist working with the corporation suggests that by using this method they could become a carbon-neutral company.[12] Using mine tailings for carbon removal would be one way to reduce emissions from mining while learning more about how weathering works.

Enhanced weathering also has a tiny grassroots following, under the rubric of "remineralization." The organization Remineralize the Earth was founded in 1986 as a network newsletter, evolving into a magazine in 1991 and eventually becoming a nonprofit in 1995. The movement has roots in agroecology, but can also be traced to the work of a German nutritional biochemist, Julius Hensel, who wrote *Bread from Stones* in the 1880s. The technology didn't exist to grind rock dust for soil remineralization at that time. But in the last few decades, some European rock companies have begun research into it again, and Remineralize the Earth draws from their work as well as the current climate crisis. They position soil remineralization as part of a new paradigm in agriculture: "The agenda for SR is clear: to create abundance in an era of diminishing resources and lead us away from fossil fuels. Remineralization is nature's way to regenerate soils. We can return the Earth to earlier interglacial Eden-like conditions through appropriate technology."[13]

Whether Edenic dream or industrial mining hell, two challenges are clear. One is that most people don't know anything about this idea. Geology is not an everyday subject, and the writing on enhancing mineral carbonization as a carbon removal strategy is dense. The second challenge is that we don't yet know enough to assess these ideas. The estimates for enhanced weathering's potential are so variable that they are practically useless, in my view, though scientists may disagree. The estimated implementation costs are $60–600 trillion dollars "for mining, grinding and transportation, assuming no technological innovation, with similar associated additional costs for distribution"[14]—a variation so wide that it strikes me as nearly meaningless.

I'm not bashing these technologies simply because they carry unknowns: they hold a lot of intriguing possibilities. For instance, enhanced weathering makes water more alkaline. Could this make it useful for reducing ocean acidification in coral reef regions? One study indicates that it could reverse ocean acidification under a low- to mid-range climate change scenario and restore global mean surface pH by 2100.[15] That would be a pretty big deal, and worth investing in. Another new paper, among the weirdest peer-reviewed science I've read, suggests that life in the seabed could make coastal ecosystems a good place to try enhanced weathering. In this scenario, microorganisms and invertebrate fauna could act as weathering agents as part of a "benthic weathering engine." Long filamentous microbes called "cable bacteria" transport electrons, making the top few centimeters of the coastal sediment acidic, and this causes carbonates to dissolve faster. There are also areas where the entire top fifteen centimeters of sediment passes through the guts of large-deposit feeders several times per year, and these "gut transits" can increase dissolution rates.[16] Potentially, the approach constitutes yet another way that life could do the work of rebalancing our carbon burden—but it's all at a very conceptual stage, right now.

I ask West what he thinks is important for people to know about weathering, and he muses for a second. "I think it's important to recognize that it's a natural process that's happening all the time, but that it's happening very slowly, and that there is this potential to make it happen faster. But then I think the other thing that's important … and I think I would say the same is true for any of the other carbon dioxide removal—to actually implement [weathering] in a meaningful way, in my personal opinion, would be enormously challenging, and so I guess it comes back to the moral-hazard question that I raised at the beginning." People shouldn't view discussion of these things as meaning that we have a technology we can rely on.

"Scientists are working on these things because we've got to turn over every stone, and maybe if we combine many different technologies, they can start to play a meaningful role. But like I said, I think it was very sobering for me to do the calculation of what the scale is that would be required for something like this. And it is maybe not technologically impossible, but it seems incredibly daunting."

This is the case with virtually all the technologies we've explored: it's not technologically impossible to scale them up, but it's daunting, for a whole host of reasons. Some of the parameters, like thermodynamics, are fixed. Some are a bit more malleable, like the qualities of plants. What could make the most difference, and really give these approaches wings, are changes in our politics, economics, and culture. In Part III of this book, we'll speculate on what a society dedicated to carbon removal might be like.

Sketch: Mountain

The peridot earrings glint green in the light. I'm holding them up when Christa calls.

"Mom?" she asks. "Are you ready?"

"Almost. I received the ticket. Or invitation. Pass? Whatever you call the thing that lets me get into the reception. And those earrings you sent. Your assistant was very kind."

"Okay, that's good," she replies. A long pause.

"What's wrong?"

"New York. They've just announced. They're building a mountain too. A hundred meters taller than ours. That's the part the press keeps repeating. Nothing about the mountain's concept, or the carbon sequestration."

"Well, the one in Qingdao is also taller."

"Sure, but everyone expects that one to be big. That's different. And they didn't have to announce it tonight."

I thread a long silver earring through the second hole in my ear, and press back my gray hair. "Christa, you told me it's all about locking away carbon. Free your ego!"

She's still quiet.

"You know it in your heart, size doesn't matter. Just think about your father. You think I would have hooked up with him if size mattered?"

"Ew. Just—no. I'm sorry I called you. I'll see you tonight." She ends the call. Another failed conversation. But she should be on top of the world. It's opening night.

I wrap myself in a cashmere cape, pull on my black boots, and wait by the open window. There's a slight breeze coming off the ocean. Over the past four years, I've been watching the mountain rise from this window. It came up over the rooftops, and then it greened up, slowly. Tonight, it's lit up for the opening, twinkling with fairy lights against the cobalt sky. It's a bit silly not to hold the opening during the day. But it is, as Christa insists, an artwork. Which connotes an evening opening. I guess.

My own mother was against the mountain. Ironically, it was the opposition of people like her to building higher that even got the mountain ballot measure passed. The city had failed to create enough housing, trying to preserve its "character," which was ironic given that the character had pretty much eroded to dust by then. Thousands of people were literally living on the streets. Housing was the compromise for Christa's art project: the mountain would have to have one stable face. Apartments on the city-facing slope would increase the city's housing stock by 23 percent, avoiding razing more old neighborhoods for high-rises. Christa offered to arrange for me to have one of the mountain flats, but I said no. I'd rather keep my tiny old studio than live on her very nice mountain.

The car glides up to the curb. I climb in and nod to the other passengers. A spiderweb of light traces the route: seventeen minutes.

"Where are you off to, all dressed up?" the woman beside me asks.

"I'm going to the mountain."

"Oh, that's right. The unveiling is tonight, isn't it." She laughs. "Not any real way to veil it, though."

"What do you think about the mountain?" I ask her.

"I voted for it. Better than just dumping all those rocks into the ocean. If the carbon fund pays for it, why not? I'm going to take my grandkids to the park next week to have a look. Maybe climb all the way to the top."

The glowing tent is set in a garden. Slate stone, jasmine air. I head immediately for the bar. The Chardonnay is a preview of the variety they're planning to grow on the mountain, the server tells me. Biodynamically, in the mist. The soil of this mountain will produce a most unique flavor, he explains. I nod, taking a step back. I notice the music start to creep in, upbeat. The lighting is warm. People are buzzing all around: big names in the carbontech / land art scene, I guess.

A man comes over: a journalist, wanting to interview me. I sigh. "How do you like the mountain?" he asks.

"Haven't been yet." Christa invited me to come during the creation, when it was just a dusty construction site, but I told her I'd wait.

"Did you always know your daughter was going to do something big?"

I have been asked this question before. What I am I supposed to say—she was always into blocks? She built fabulous sandcastles? She drew great pictures of mountains? "I knew she was a visionary. And that she was persistent enough to stick to a vision, and that she could collaborate with others, and charm them into her vision. But no, I didn't know she was going to generate a mountain." I can tell he likes this quote. I excuse myself to go look at the exhibition. There are full-color photos from the mountain's construction, as well as an interactive display about the coral reef and aquarium beneath its slopes. I skip past the materials science part; I already know all about the underpinning structure and the technical feats of building the skeleton and casing—how only the parts that are designed to weather will crumble into the sea lapping at its edges. More interesting to me are the native gardens on the ocean-facing side, and the way they taper off into a blank space, where it remains to be seen what will take hold in the gray-green sand. This was the most audacious part of the design, and the part that Christa had to fight for the hardest. She wanted a whole face of the mountain to "design itself," as she put it. *I want the viewer to experience and aid in the weathering, to contemplate geologic time. For their feet to send off mineral grains into the sea.*

The room is getting quiet. Heads turn. Christa and her entourage are electric, beaming. Someone hands her a microphone.

"I'm so grateful you all could be here tonight," she begins. She talks about all the teams—the engineers, the hydrologists, the landscape ecologists and wildlife biologists, the housing architects, the museum designers, the policy and regulation team. My attention has wandered by the time she gets to thanking the lawyers; I need another glass of wine. "And I'd like to thank my mother. Mom, where are you?" I straighten up, smile. "Throughout all of this, I knew my mother had my back. I'm so excited to finally share this mountain with her today. Mom, you always believed in me, and I couldn't have done this without you." An involuntary lump swells in my throat. Everyone around me is clapping. She gestures me forward. I wave, no thank you, but she insists; I make my way through the fancy dresses and bright lights, and give her a hug. She's got dark circles under her eyes, but they're sparkling like they did when she was a child.

"Let's see this mountain," I tell her.

Part III: The After-Zero Society

6

Working

What would it be like to live in a society that's brought greenhouse gas emissions not just down to zero, but into negative territory? Who works there? How did they learn the skills they need to know? And what are their rituals, their aesthetics, their emotional lives like? Do they pride themselves for having come down a curve of carbon dioxide emissions and temperatures? Does success in achieving the feat of climate restoration define aspects of their culture? Or is that curve long forgotten?

Exploring an after-zero society means playing with utopian possibilities. These may seem like flights of fancy in today's world, where men of industry are making presentations about "carbon negative" oil for "negative emissions," preparing to capture the concept of carbon removal to prolong the life of fossil fuels. Even so, it's worth taking a moment for utopian exploration. The casual dismissal of utopian thought is linked to an oppressive politics. Marxist feminist scholar Kathi Weeks observes that "political realism tends to be associated with a mode of hard-nosed, hard-ball politics," while "utopianism can be understood—building on this traditional gender logic—as both softhearted and softheaded, or, more precisely, softheaded because softhearted."[1] Social relations are stabilized by claims about their natural basis—for example, claims about how women "naturally" are—and analyses that propose alternatives are often dismissed as unrealistic, Weeks writes. It was for this reason that the eighteenth-century feminist writer Mary Wollstonecraft was forced to say that even her moderate visions of gender equality could "be termed Utopian dreams."

When you dig deep into the discourse of what's possible with regard to climate change, you find similar claims about the "natural" state of things—that it's human nature to degrade our environment, for example, or that humans will always place their own and their group's self-interest first. That humans won't choose to do things unless they garner economic incentives. And to suggest otherwise will inevitably garner labels of utopianism.

As science fiction author Kim Stanley Robinson writes, it may be easy to imagine a radically different society, "in that one merely expresses wishes and defines some version of justice, equality, peace. That's all easy. What's hard is imagining any plausible way of getting from here to there."[2] Robinson writes that perhaps when Marxist theorist Fredric Jameson talks about the future as being "unimaginable," he doesn't mean that the future is an unimaginable destination; what's unimaginable is "a history to a good future place." Trying to imagine it anyway, though, is valuable, Robinson says: doing so points to the problem, and it generates new stories.

So, is this after-zero society techno-utopia? Or is it a small-is-beautiful utopia? Perhaps it can be both. I'm interested in a synthesis between these industrial technologies and something that appears to be on the other side of a binary from them: degrowth.

Environmental scientist Giorgios Kallis defines sustainable degrowth as "an equitable downscaling of production and consumption that increases human well-being and enhances ecological conditions."[3] The organizing principles are simplicity, conviviality, and sharing. Note that these are *organizing* principles, not values (though they are probably that, too). In a degrowth mindset, innovation is "directed towards new social and technical arrangements that will enable a convivial and frugal living." Yet the view of many degrowth advocates is that technologically complex systems beget technocratic elites: fossil fuels and nuclear power are dangerous because sophisticated technological systems managed by bureaucrats will gradually become less democratic and egalitarian.[4] Large-scale technological systems, the argument goes, result in a society divided into experts and users. This limits the engagement of degrowth thinking with many forms of carbon removal, which is unfortunate. Many tenets of degrowth encompass what I think a best-case scenario of carbon removal would look like: directing innovation toward conviviality, frugality, and also justice. As Kallis sees it, "sustainable degrowth denotes an intentional process of a smooth and 'prosperous way down,' through a range of social, environmental, and economic policies and institutions, orchestrated to guarantee that while production and consumption decline, human welfare improves and is more equally distributed."[5] The "way down" dovetails with the narrative of removing emissions. Some of the proposals Kallis sets out in his book *In Defense of Degrowth* include basic and maximum income, green tax reform, cessation of subsidies for pollution and reallocation of these funds toward clean production, support for a solidarity society, optimization of the use of buildings, and abolition of the use of GDP as an indicator of economic progress. Many of these would enable carbon removal at scale.

Whatever its particular form, what's clear is that we need a social imagination to match our technological imagination. And these perspectives of degrowth and ecomodernism do not comprise two ends of a continuum. For we will need more of certain kinds of industries, and less of others, and we can use a lot of the same tools that degrowth advocates are calling for toward many of the ends degrowth theorists favor. Reiterating a tired binary between ecomodernism and withdrawal makes it impossible to see what we need to do. Critical thinkers have been so focused on documenting the unfolding crisis that we don't focus on the ways in which industrialization could advance. But there may, in fact, be a hybrid position, a third, distinct frame. What about a democratically controlled industrial technology that doesn't exist to "conquer" nature?

There's clearly no single answer to what an after-zero society looks and feels like, who lives there, and what they will value and do—but there are manifest interesting possibilities. This terrain is made more complex by the variety of technological and non-technological trends we might anticipate over the next century.

Some of them we can reasonably foresee: demographic shifts to a population of 9 or 10 billion, advances in machine learning and synthetic biology, transformations in the nature of work and education, and climate change itself. Other trends will be surprises. However, I'm looking at this terrain with an eye to how a culture committed to carbon removal would interact with these trends, and backcasting from there.

~

Work in the twenty-first century is complicated. The shifting backdrop includes stagnation of wages and rising inequality. There's automation and machine learning, and the jobless class they portend. There's the cult of the entrepreneur, posited as a savior from these trends. Trade wars are started over the ideal of domestic manufacturing jobs; people are trying to figure out what "dematerialization" means and if it even exists. Conversations are spinning in public spaces about race, gender, and employment. Academic venues are abuzz with discussions of how nature gets enrolled to do work for us. And there are billions of people subsisting on a few dollars a day, many of whom are forecast to move from rural regions to cities in search of work, where they may find only informal employment. Meanwhile, people in industrialized countries are working out what kinds of livelihoods will be had in emptying rural villages and towns.

That's a lot going on. Suppose, emerging from all of this preexisting context, that there is also a massive effort to remove carbon from the atmosphere over the course of this century. Who does the work of carbon removal? What sort of work will it be? Will these be "good green jobs," creative jobs, hard labor? Who will the employers be: the state, startups, reimagined large firms, tech companies, cooperatives, local organizations? Will the burden of jobs fall disproportionately upon one segment of society—or will they be exclusively granted to one privileged group?

I'd like to begin this inquiry with a (true) story about what was reported in the press as a geoengineering event—one of the only "geoengineering" events ever to happen. It took place in 2012, a few hundred kilometers offshore from Haida Gwaii, which is a remote archipelago near northern British Columbia. A chartered fishing vessel had poured iron into the ocean in order to generate a plankton bloom. The branding of "rogue geoengineering" crystallized quickly around the endeavor; a staff writer in the *New Yorker* called it an act by "the world's first geo-vigilante."[6] From another point of view, though, some narrators saw the story in terms of beginning the work of climate and ecosystem repair when no one else was stepping up to the job.

Taking action

In Old Massett, the rain drives hard across vacant storefronts. A hand-painted plywood sign hangs off a shuttered building: True North Strong and Tanker Free. The cannery and docks are quiet; stacks of crab pots corrode. "Notice to clam diggers: The legal minimum size for razor clams is 4 inches." Gravel lots host abandoned shipping containers, rust upon gray. Flocks of ravens swoop down to devour bright-red salmon scraps, scattered along the rocky shores. In the shelter of a café, a flier is pinned to a bulletin board: "British Columbia Reconciliation Week: Be Part of the National Journey for Healing and Reconciliation!" Bullet points include: "Statement Gathering," "Traditional Ceremonies," "Survivor Gatherings," "Education Day," "Witnessing Survivor Statements," "Cultural Performances," and "Films and more."

Old Massett, to an outsider on a rainy day, feels desolate, a village of 600 souls that is quietly struggling to survive, to stay intact. Unemployment might reach 70 percent, though this is mitigated somewhat by people's ability to fish and gather traditional foods. I talk to two guys on the wharf; they tell me the best field for wild strawberries is behind the school. The season for coho is ending in two weeks. "We're natives, so we can catch our own fish," they tell me—all they need is potatoes and rice. They freeze thimbleberries for smoothies. But the village is in a tough place. The most significant industries here are fishing and logging, and the fish catches vary, with licenses often granted to big companies from the main-land. If they manage to pick up enough shifts in the fish-freezing plant, they can get employment insurance assistance—$860 Canadian a month. But right now, the cannery is quiet.

There are about 2,500 indigenous Haida people in the Old Massett band, and another 2,500 in the band of Skidegate, about an hour and half's drive south. Once, of course, there were many more Haida villages. After contact, the population decreased by between 90 and 97 percent. The Haida spent the following decades being forced into residential schools, their language banned and then endangered. Colonizers came to cut their forests and deplete their fisheries. But to write this is not to make Haida people out to be passive agents in the colonialization of their lands: their resistance to logging forced important changes in Canadian law. They have launched fishery and stream restoration projects. They are forward looking and creative.

The image of people resisting the imposition of someone else's vision for their land and climate is familiar: we know the defiant faces at protests; the aerial photos of marchers, holding signs that say no to things. There is a certain comfort in these captioned photos: 1960s nostalgia, the idea that people are still out there resisting, that the resistance means some kind of democratic deliberation is happening. Resistance to environmental degradation is scripted, expected. Actively pursuing environmental interventions is less so.

The dependence of people on Haida Gwaii upon on the land makes it all the more important that they have access to local foods, like salmon, and to functioning ecosystems. In a place where the environment is cherished and relied upon, and where there's also significant economic hardship, the idea of payment for ecosystem services takes on a different rationale than the get-rich-quick exuberance of financialized carbon. This context is important for understanding why leaders in the village of Old Massett decided to finance an intervention aimed at restoring salmon populations, and possibly sequestering carbon, in conjunction with an outside entrepreneur named Russ George.

The event happened as follows: In the summer of 2012, a small crew took a thirty-five-foot rented fishing vessel, the *Ocean Pearl*, out onto the blustery seas far beyond Haida Gwaii, in an area considered to be Haida territorial waters. The crew poured amounts of iron into the ocean over the course of a few weeks at sea, and they remotely monitored the resultant plankton bloom with two ocean gliders and twenty drifter robots. "We added approximately forty kilograms of iron-based material per square kilometer of ocean and thereby changed the background concentrations of iron in the top one hundred meters of ocean from approximately three parts per trillion to ten parts per trillion," reported the former scientific director of the Haida Salmon Restoration Corporation (HSRC), Jason McNamee, when I talked to him some years ago at their office, then in Vancouver (some time before it was investigated by the Canadian government). This event, however, was seen by many not as stewardship, but as a harmful intervention. "We never did and do not consider it geoengineering," McNamee said.

Geoengineering or not —why, of all things, would this corporation try ocean iron fertilization? And why would a band of environmentally attuned indigenous leaders agree to fund it?

Here's why ocean fertilization, or adding nutrients to the ocean to "fertilize" it, is interesting for carbon sequestration. Phytoplankton —those tiny little bits of living matter in the ocean—are responsible for half of earth's primary productivity. There's a lot of biomass growing and churning and dying in the oceans. If more of it can grow, and sink to the depths of the ocean, it could in theory store a significant amount of carbon. Essentially, it's a way of biologically enhancing the ocean sink. What limits plankton from growing, in certain areas of the ocean, is a lack of nutrients. So: more nutrients (fertilizer), more plankton, more carbon sequestered. Broad strokes, it sounds like it has a lot of potential. But the fine print: we don't know that much about the actual carbon sequestration potential of this method, because for it to be truly sequestered, it has to sink down to the deep ocean. It's fantastically difficult to measure and study this; the deep ocean is like another planet. Sequestration also depends upon the availability of light, silicate, and other factors—and there are potential side effects. Because of these mea-surement difficulties, and unknowns, it would be really tough to actually set up a market for this kind of intervention.

However, this wasn't the only motive named by the HSRC. As their name might suggest, they also saw ocean iron fertilization as a way to accomplish salmon restoration. The Haida, who have operated a successful hatchery on the Yakoun River for over forty years, know that salmon go out into the deep ocean. After they leave for the deep ocean, however, no one knows exactly what happens; what *is* known is that, although salmon runs fluctuate wildly, they are not coming back to Haida Gwaii in large numbers. Moreover, as indigenous scholar-activist Kyle Whyte has pointed out: "The declining salmon runs do not arise only or primarily from the looped back effects of recent anthropogenic climate change. They are due to factors including land dispossession, disrespect of rights and eco-logical degradation."[7]

The HSRC wanted to do their own research on this matter. They were working with a theory that has to do with dust, something that the entrepreneur Russ George also talked a lot about. The idea goes like this: Phytoplankton are in decline. Growth of plankton in the North Pacific is known to be limited by the amount of iron in the water. The fantastic 2010 salmon run could be linked to the explo-sion of Kasatochi volcano in Alaska, which rained nutrient-laden ash on waters. The logic was thus: more dust leads to more plankton, leading to more creatures in the food chain, which ultimately leads to more salmon. This is a striking feat of cognitive linkage—one that would be extremely difficult, maybe impossible, to "prove" according to scientific standards. It would likely be thrown out in a research grant proposal because it would be so difficult to gather robust evidence for making these kinds of causal claims.

But if you're not a scientist at a grant-funded research institution —if your interest is applied, and your product is not citations and CV line items but increased ocean productivity and pelagic biomass— maybe offering robust, empirical proof isn't as relevant, because if it works, it works. The project was dubbed a "nonscientific event" by Environment Canada; but to truly understand it, it's more useful to think of it as a *different type* of science; perhaps "free range" science, following environmental geographer Rebecca Lave.[8] The methods, funding sources, and intents were often a hybrid of conventional and unconventional, or professional and amateur, practices. The project was conceived by people who didn't have advanced training in science, but wanted to do the work.

Much of the condemnation of the project as geoengineering was linked to Russ George, a familiar character to activists concerned about ocean iron fertilization. George has a checkered past with carbon credit startups. His earlier company Planktos was chased out of the Galápagos following public outcry back in 2007, with George defending himself as a "first responder to a planetary medical emergency," claiming the terrain of responsibility for restoring nature.[9] Outrage over his proposal in turn helped move forward the 2008 Convention on Biological Diversity moratorium on large-scale commercial ocean fertilization. George makes an eccentric villain. Yet focusing on him as the "rogue geoengineer" ignores a lot of important contextual facets of the 2012 HSRC project. For one, the village has a long-standing concern with both climate mitigation and adaptation. The economic development officer who worked with the salmon restoration project, John Disney, also had a vision for comprehensive stewardship that included transitioning to independent wind and tidal energy, strengthening the local food system to become more sufficient, and reforming health and transportation systems. For example, years after this ocean fertilization project, he helped get a new biomass boiler built to heat community buildings, replacing polluting diesel with electricity generated from wood waste. Also among his past efforts was an attempt to generate revenue for afforestation through carbon credits, from which he learned that "one-off" projects aren't sustainable. This is, in short, because no one is going to pay communities to conserve their resources. What happened with this ocean fertilization project, though, is that the village voted to use their own money. As McNamee said, the Haida of Old Massett Village "have a cultural imperative to steward their resources," and since efforts to curb global emissions have been unsuccessful, we have a responsibility to look at alternative solutions.

Changing the earth is still often imagined in Cold War terms. We are still under the spell of images of mechanization, desensitization, and compartmentalization, and they tint how we think about engineering the climate. It's supposed to be people in drab clothes who are scheming up climate control—throwbacks to imagined bureaucratic, monocultural hierarchies—or else a constellation of conspiratorial businessmen. Not indigenous people—a fisherman working and living in a tiny village, or an elected councillor who admits to an awkward media presence. "Geoengineering" wasn't supposed to be a village project, and the script of "geoengineering" makes it hard to understand this project in its context. The script calls for a confrontation between actors who haven't even shown up for the performance.

Still, some of the characters in how this story is usually told are contemporary enough: the entrepreneur and the eco-hacker; the eco-startup or "ocean biotechnology and stewardship corporation." Who these days gets to be a "maker"? Or a "decision maker"? Plenty of climate engineering startups have popped up (and some have already gone silent): Climate Engineering (industrial air carbon capture), Global Thermostat, Kilimanjaro Energy, Biorecro (bioenergy with carbon capture and storage), Cool Planet (biochar), to name a few.

People who are taking action, being creative, or disrupting often have affinities with this startup culture. Silicon Valley's cultural roots have been traced, most notably by media scholar Fred Turner, both to the collaborative, interdisciplinary, cybernetics-infused work being done in federal labs during the 1940s and '50s, as well as to 1960s counterculture—specifically, a tech-friendly strain of this counterculture, filled with ideas of holistic systems ecology and back-to-the-land New Communalism.[10] Russ George fits in perfectly with such a cowboy-nomad archetype, and the press picked up on this. Rogue, maverick, wrangler of credits on the carbon frontier, guest speaker at Silicon Valley's Seasteading Institute, where libertarian tech gurus plan their floating existence loosed from the shores of the backward-leaning mainland. And his emphasis on "ocean pastures" and "ocean ranching" extends the cowboy archetype more literally. Press treatment of George reflects the media's love-hate relationship with the entrepreneur and the hacker, the ones who are rolling up their sleeves and playing around and *doing*, rather than theorizing. Startups get to do the unconventional work, something for which they are distantly admired. But we're also skeptical of their sense of responsibility, and their drive for disruption.

Even though George is a solo entrepreneur, there is also a conversation here about what kind of jobs can be created for whole communities, especially in rural areas. Jason McNamee dreamed of taking collaborative science to the next level with a Haida Ocean Center of Excellence to study marine changes: "We would also aim to innovate new ways to study the ocean and empower citizen science by designing and building cheap DIY instruments, the plans and software of which would be publicly available," he told me. "Such a center would require shore-based, ship-based, and robotic infrastructure." HSRC was already able to borrow needed equipment and get input from professional scientists—many of whom weren't interested in formal association with the project, but *were* interested in the data. Science has been set free from the big institutional laboratory, and the processing power and tools to gather and handle large datasets are increasingly affordable to a broader range of people. The cautionary downside to this kind of research is that it may miss key bits of information or broader understanding, the kind one gets from years of training and layers of institutional oversight. Yet if scientists began to collaborate with citizens, amateurs, or people who didn't spend years as underpaid graduate students, perhaps this would be a great democratic shift, and good for science. Perhaps it would mean that people beyond big cities and big laboratories would get to participate in the knowledge and information economies. If carbon removal can be done in places beyond big cities—in a robust and permanent way—those projects could help support knowledge-economy stewardship jobs in rural communities.

In what follows, we'll pick up on some of the threads in this story: the plight of rural or extractive economies, the roles of the hacker and the disruptor, the robotic ocean gliders, and the plankton itself.

Rural labor: Burden, or carbon care work?

The transition toward a society dedicated to carbon removal at climate-significant scales could be an opportunity for rural reinvigoration—or conversely, one for rural oppression and the continued transfer of wealth out of rural lands. This goes for both developed and developing world contexts, though each bears its own challenges. In either case, we should see carbon removal through the lens of *rural* economic development issues. When it comes to soil carbon and regenerative agriculture, revitalization of rural economies is certainly already included in the conversation. Yet when it comes to high-level policy, or research on "negative emissions" or carbon removal writ large, consideration for how these technologies will intersect with rural communities is conspicuously absent. Perhaps this dynamic is omitted because most people talking about carbon removal live in cities, or because they don't readily associate "technology" with "rural"—another persistent binary.

There are tremendous opportunities for rural communities to benefit from carbon removal practices based in cultivation and burial. On the cultivation side, carbon removal policy could provide economic opportunities for farmers who take up regenerative agriculture. Public awareness and support from urban food consumers and taxpayers is crucial to bridge this gap between targets and reality, but so is demand from rural areas for subsidy redirection. On the infrastructure side, building out infrastructure and operations for carbon capture and storage with direct air capture or bioenergy would offer jobs with a similar skill set to those in the oil and gas industry, provided an effort were made to retrain workers into this field. Enhanced mineral weathering of mine tailings for fertilizer would offer mining and transport jobs. Connecting with unions and workers would be key here, and carbon removal could become part of a just transition away from fossil fuels. In any case, carbon removal would be more likely to succeed and actually reach net-negative emissions if it emerged in dialogue with what rural communities need.

There's a funny tension here: Is this work a burden, or is it a privilege? It depends. Take carbon farming: on one hand, it entails a bundle of practices that entail various new burdens. Farmers need to purchase a different set of equipment, and it can take several years to make that transition. As a side effect, farmers are also cleaning up someone else's waste. Growers, on the other hand, get to be landscape designers, exerting control over land and resources. If you're a day laborer or hired farmhand, though, you're executing someone else's vision, presumably with little say. This brings up the question of collective labor and democratic decision making in farm design and operations—something recognized by the principles laid out in the "Regenerative Organic" certification scheme. It has "Social Fairness" as one of its three pillars, with several standards for "Farmer and Worker Fairness" around a living wage, transparency, and the right to form associations.[11]

When we think of environmental maintenance work exclusively as a burden, we miss something important about what motivates people to intervene, to take action, to do this caring work. In current environmental policymaking, writes development scholar Neera Singh, conservation is regarded as a burden that entails high opportunity costs, and so financial incentives come to be seen as critical to offsetting those costs.[12] But in fact, they aren't necessarily so critical. Singh has studied community forests in Odisha, India, for twenty years. What she has found is that when payments are established for ecosystem services, the payments are often insufficient to compensate for lost income and opportunities. Rather, efforts to conserve forested land depend upon a host of nonmonetary, personal, and collective motives, including sacred values and intergenerational concerns. Conservation care is described by Singh as "affective labor," and as a "gift." Affective labor, in contrast to alienated labor, writes Singh, involves self-expression; its ideal is a craftsman or artist, who expresses their inner self and gives to society as a whole. It can't be separated from the person doing it. The relationship between the person and the object of labor is crucial.

In the "gift" paradigm, people act not as buyers and sellers of environmental services, "but as reciprocal partners who share both the burden and joy of environmental care." Singh cautions that "caring labor framed in the language of a gift can be potentially exploited, as women's caring labor continues to be." But, she explains, this framework also means that markets and purchasing power are no longer the arbiter of how resources are allocated: power is shifted to the gift givers, whose power comes from the reciprocity inherent in the gift. One challenge with a gift paradigm when applied to carbon management, as environmental humanities scholar Karen Pinkus points out, is that the recipients are humans in the future: "On first glance we could say that the recipient cannot give back to the giver because they do not occupy the same space/time except in the most phantasmatic sense. Wouldn't carbon management, then, overcome the temporal aporia that makes the gift impossible?"[13] In this sense, perhaps carbon removal implies a greater ask of the gift: a gift that will go unreciprocated during this generation.

The takeaway here is that with carbon removal, just as with land stewardship, initial payments might be insufficient to motivate action. "Incentivization" is the go-to way of thinking about how carbon removal would be accomplished; and it will likely be very important, but it's not the only motivator. Have we been brainwashed into thinking that paying people is the only way to get things done? It's relatively easy to imagine farmers taking the perspective of affective labor, stewardship, and the gift—many growers I've talked with certainly speak a form of this language, though it doesn't translate well to conventional farm operations. How could this perspective apply to industrial carbon capture and removal? A pipe fitter, for example, might very well view their work with pride, as something that powers the nation. The failure by urban elites to recognize their labor as such contributes to the divisions in today's society.

It's possible to simultaneously hold all these perspectives as true—carbon removal as burden, opportunity for work, and form of care—while at the same time not losing sight of the dimensions of responsibility (who should be responding to this duty) and agency (who has the capacity to do so).

Tech labor: Coders, makers, and entrepreneurs

"In the future, carbon removal is going to be automatic, and simple, and happen invisibly in the background," says Paul Gambill, CEO of Nori. Nori is a startup interested in disrupting a broken system: carbon markets. The aim of the company is to use blockchain technology to set up a voluntary marketplace that will connect people who will pay for carbon removal and people who can actually remove carbon. The gift cards, or tokens, are used to pay for carbon removal credits. Gambill is explaining all this to a room full of carbon removal enthusiasts at an event called Reversapalooza—"a collaborative summit for climate change reversal." He invites us to imagine that we're in a Lyft or Uber. After we get out, we receive an alert asking us if we want to have a sponsor pay to negate the emissions of that ride. "Or imagine that you drive your car up to a gas pump, and after you fill it, in the same transaction, you can pay to purchase carbon removal certificates that will negate the emissions from that tank of gas." Or imagine a mobile phone game about environmental restoration, in which in-app purchases go to carbon removal. In this room, the vision seems possible.

Outside, it's a bright, blue-sky spring morning on Seattle's waterfront. Sailboats from around the world are gathered for the Clipper Round the World race, colorful flags fluttering from their masts. Ferries are crisscrossing the shining harbor, framed by the snowcapped Olympic Mountains on the horizon. And in a ballroom in the basement of the Marriott Waterfront, a hundred people are playing a game in order to better understand how blockchain can help reverse climate change. Soil scientists, blockchain enthusiasts, activists, and climate foundation professionals are roaming around, making transactions. In this game, buyers of carbon removal credits have the goal of offsetting our emissions by buying carbon removal credits. Sellers are trying to generate additional revenue by selling these credits. My tablemate and I are sellers, simulating a vineyard enterprise that has the potential to remove 1.6 million tons of CO_2 during year one—but it'll cost us 1.9\$ million over two years. The black hotel tablecloths are stacked with carbon removal credits and tokens representing the eponymous Nori cryptocurrency.

To better grasp how blockchain can be used in carbon removal, I spoke with one of Nori's founders, Christophe Jospe. Nori embraces Buckminster Fuller's words of advice: "To change something, build a new model that makes the existing model obsolete." "We're trying to say, this whole top-down approach of Paris isn't working," Jospe explains. "They continue to change the numbers around so that their voluntary commitments will work. But if anyone actually does the math, you'll see how much of a scam it is—let alone how much of a scam it is that all these countries are actually lying on their greenhouse gas reporting. So, screw that; let's start a voluntary marketplace with people who just want to pay to prove that carbon removal works."

Carbon markets have been set up in ways that destine them for failure. For one, there's rampant double-counting. Currently, what often happens is that when a carbon polluter in California buys a credit from, say, British Columbia, the emissions reduction is counted both in British Columbia (where it really happened) and in California—an accounting mess that ultimately makes the system dysfunctional. If you sell someone a good, then that good can't be in two places at once—but right now it's kind of working that way. Article 6 of the Paris Agreement addresses double-counting, stating that parties should follow UNFCCC accounting methods and "shall apply robust accounting to ensure, inter alia, the avoidance of double counting." In other words, if you're a country selling a credit, you can't simultaneously count that credit on your emissions reductions. However, this only applies to parties in the convention (not corporate buyers, for example), and it still seems a bit shaky and unresolved. On top of this killer flaw, there's the fact that methodologies for carbon accounting are often proprietary. And then there are all kinds of middlemen that increase the cost of carbon credits, making it harder for smaller actors, including smaller farmers without a lot of initial capital, to participate in these markets. In short, the whole broken carbon market system is in need of a redo.

Nori's starting marketplace is filled with regenerative approaches,

Jospe says, and they're working with farmers to build a software platform that will make their carbon accounting happen out in the open. One key feature of the blockchain approach is that that the token retires the carbon removal credit immediately, so it can't be resold. "And that fixes a lot of problems that are happening in today's environmental markets, right off the bat."

If you're not too familiar with blockchain, it's important to know that it is not only a way of setting up alternative currencies. It's also, essentially, a distributed ledger, a virtual notebook of transactions that's not held in one place, but is available all over. As London-based urbanist Adam Greenfield writes in *Radical Technologies: The Design of Everyday Life*, there aren't really any handy metaphors for understanding how blockchain works to create and mediate a means for the transmission of value—how does a chain of signatures become a coin?[14] Here are the essentials: The blockchain is a digital medium of exchange. Take, for instance, a routine digital transaction with which you're familiar. You buy a cup of coffee, pay with a debit card, and three dollars is debited from your account. This means that you trust the bank on the other end not to subtract thirty dollars instead, and it also means that you're not using that same three dollars to pay for a croissant, too. You trust that the bank is handling things—the bank is central to keeping the record, the ledger.

Right now, the carbon markets don't have that trusty ledger, the bank-like entity that efficiently keeps those three dollars from being spent simultaneously on both the coffee and the croissant. What a cryptocurrency solution implies is: let's get rid of the central entity keeping the ledger—the record of transactions—and inscribe it instead in the token of carbon removal itself. A blockchain is a shared public record of transactions. The coin has a signature, and each transaction has a unique record, which, with a currency like Bitcoin, is propagated through the network in an encrypted form. These records get stacked into blocks (figuratively), which then appear in a chain. Why is this better than having a master international carbon bank keeping records? One reason is that it cuts out the bureaucracy and the increased costs of administration. A more fundamental appeal is that with blockchain, no single entity holds an oversized amount of power—the distributed system is a more robust, secure system. You don't have to depend on some external entity to guarantee and take care of your transactions. Greenfield lays out the promise of blockchain beyond Bitcoin: the transparent exchange of any fiduciary token from one party to another, including assets, liabilities, obligations, wagers, votes, and so forth. Greenfield explains:

> This prospect opened up extraordinary possibilities for the world of administration, which in large part consists of little more than keeping track of such positions as the information encoding them is moved through, across, and between organizations. More profoundly yet, it presented new ways of thinking about organization itself—about what it means to associate with others, how joint intention might be harnessed, and parties unknown to one another yoked in effective collaboration, across all the usual barriers of space and time.[15]

For a joint intention that needs harnessing, like carbon removal, blockchain could be a significant innovation. (People who've heard about the enormous carbon footprint of Bitcoin ask how this is possible, but again: blockchain is more than just Bitcoin. Bitcoin has become known for the tremendous energy consumed by its "proof-of-work" computing needed to encrypt transactions and mine new bitcoins. Nori uses Ethereum, a cryptocurrency that is switching to a "proof of stake" system that requires less electricity.) A blockchain system could address the double-trading problem, and it provides a way to deal with the trust issues around shady carbon removal schemes. It offers the possibility of smart contracts—contracts that are only executed when certain conditions are met, such as verifiable storage. At its most expansive, blockchain offers an orchestration architecture for getting millions of actors who don't know or trust each other working toward a global goal.

But when it comes to carbon removal, what's being administered is really quite material—not to mention invisible and hard to measure. Thus, for carbon removal cryptocurrency to work, there needs to be a way to verify that carbon is stored with some degree of permanence: the trust needs to be *grounded*, so to speak. As Nori's white paper notes, buyers in the voluntary carbon marketplace value the public relations benefits of the purchase, but they are often daunted by the complexity, opaqueness, and amount of money spent on auditing and compliance: "The most important factor in their evaluation of a carbon removal tool is how well they can trust that the carbon was actually sequestered."[16] The platform for recording and storing trustworthy data is what Jospe and his team are currently in the throes of hashing out.

Measuring carbon storage in soil, for example, could rely on overlapping layers of data: imagery from drones and satellites, combined with a sensor in the soil or a core sample. Then, permanence is assessed using probability of reemission. This data goes into the blockchain, followed by a verification process, which includes auditing all this and making sure the verifier does not collude with the farmer.

"So if one or two verifiers come, would they be like independent contractors who decide to do this for a job?" I ask.

"We want to create a framework so that verifiers can come and participate," Jospe explains, suggesting there would be white-listed, or trusted, verifiers. Open-source methodologies are another feature that would build trust. "It's ridiculous that methodologies for the carbon accounting are treated like intellectual property," Jospe explains. It's cumbersome to get a new methodology approved; and then once the method is set, it can't be copied. Jospe and colleagues believe that the methods should be open source, so that they're not owned by any one person and can improve over time.

Blockchain and the carbon market space have each been described as a Wild West, filled with hype, scams, and dishonest operators. Jospe, too, is concerned about quality and transparency. "Well, why should we trust this tech startup that's saying they're going to reverse climate change? Who are they? This cocktail and conference crowd that just spends a lot of money on airline tickets going to climate conferences around the world, saying the same thing over and over, pushes down a lot of groupthink—I think we will run up against that," he reflects. "There are companies that are writing a white paper, putting a team together, saying they're going to build all this fancy software, making a hundred million dollars, and then walking away. It's a total sham. And it's a problem, it's a real concern. That's why people are calling the technology a bubble, and it doesn't help anyone. But we're building a platform so that people who buy these gift cards can use them immediately, and it's ready to go."

Is this move toward blockchain for carbon removal inevitable? I ask Jospe.

"What's inevitable is the recording of a physical activity into a digital asset—that's going to happen, whether Nori gets it right or not. That's going to happen. And that's a real game changer," he tells me. On one hand, some readers who are familiar with the literature on the commodification of ecosystems will see this as another step in the wrong direction, toward further abstraction of value. On the other hand, blockchain for carbon removal involves trading an obligation certificate backed by a real-world action. And in this sense, there's a radical streak in cryptocurrency. "There's a lot that the world maybe hasn't caught onto yet—and this isn't me being some kind of anarchist that wants to burn down the government— but our entire monetary policy is completely screwed," Jospe states, citing the absurdity of the US government continually raising the debt ceiling, and how the massive debt accumulates. "I think a lot of the cryptocurrency idealism says, well, there's a better monetary policy in general. That's kind of a tangent, but that plays in some ways into what we're doing." Another fundamental tenet of Nori's perspective is that there should be open-source tech for carbon removal. Their Seattle meeting featured a panel on this, drawing together people working on carbon-smart building, household soil-carbon measurement devices, open-source tools to track industrial carbon value chains, and more. It was clear that open-source tech for carbon removal was something that most people in the room were in the early stages of envisioning. It was also clear that the energy in the room was much, much higher than at any other meeting in a windowless hotel ballroom that I've ever attended.

What kind of work does Jospe do, exactly? There's a software engineering component to his team, but also a visionary component.

He cohosts the delightfully acronym-free Reversing Climate Change podcast, which aims to render the nuances of carbon removal in plain English. There are aspects of design and visualization in this sector, too, as well as cultural and messaging work—producing the narrative of carbon removal. It's a knowledge economy job that interfaces with many other types of material, nonhuman, and machine labor.

The easiest descriptor of Jospe, though, is probably that of the entrepreneur. He worked in academia, but became frustrated with its bureaucracy. He worked in consulting, but devoured books on entrepreneurship and realized that he *was* an entrepreneur. The voluntary carbon marketplace was something he always wanted to work on. He likes the ability to be his own boss and create a positive work culture around him; he also explains that it comes with risk, and that he is in a privileged position to take on that risk, with no debt and no dependents.

Another entrepreneur in the carbon removal space, Tito Jankowski of the aforenamed Impossible Labs, similarly sees value in the entrepreneur's freedom to try things. In Jankowski's view, we are closing out an era that focused on scientific monitoring and scientific discovery. "[In] the era that came before, there wasn't room for mistakes. It was all about perfect spreadsheets, and charts, and models." Now, we're in an era of solution building, where entrepreneurs are needed to take a shot, to fail, to try things.

Jankowski seems to see direct air capture as an idea that is lofted on this kind of freedom: taking carbon dioxide molecules out of the air is one simple thing. "It's not, okay, let's go to the smoke stack and make a deal with them so we can take their stuff, and then we can make a deal with the government so we can get a carbon credit." Rather than getting entangled with the political world and having your efforts ground to a halt, with direct air capture, you're able to take the first step (and Jankowski's literally produced a step made from carbon captured from the air, as an artifact). The struggle is, of course, that there's no market for captured carbon. But Jankowski sees it all as an opportunity: "The technologies of carbon removal, of pulling carbon from the atmosphere itself, and learning to build a market and a use for it, to me ... this is like our ultimate opportunity to take everything that we've learned as a society ... everything from our arts, and our music, to our science and our entrepreneurs, and just saying, this isn't a bad, doomsday, everything-sucks thing." Rather, climate change is, he says, "the biggest opportunity I could have ever, ever dreamed of."

Transforming a trillion-ton existential disaster into an opportunity takes audacious thinking, but that's what entrepreneurs are encouraged to do. Of course, this also coincides with the retreat of the state under several decades of neoliberalism. Entrepreneurship has become a model of how to be and behave, as cultural theorist Imre Szeman reflects (and perhaps this is the new era that Jankowski was musing on, articulated from a different perspective). Szeman writes: "We are all entrepreneurs now, or, at a minimum, we all live in a world in which the unquestioned social value and legitimacy of entrepreneurship shapes public policy, social development, economic futures, and cultural beliefs and expectations."[17] Entrepreneurship is held up as a way to gain income and give one's life meaning, especially when many workers in the global North have "bullshit jobs," Szeman notes (following David Graeber). Generalized risk is at the heart of this: "Everyone has to be an entrepreneur because in the absence of society—of the guarantees of formal and informal security and welfare once provided by community and state policies and programs—risk is a universal condition of existence." In Szeman's analysis, "Instead of chafing and complaining about the retreat of the state and the disappearance of society, or about their abandonment to the hostile environs of the contemporary labor market, entrepreneurs embrace the openings left behind by the retreat of the state as spaces wherein they can shape their own subjectivity with the greatest freedom imaginable." While this aligns with Jankowski's comments about freedom, it's also a little bit of a caricature: an entrepreneur like Jospe can very well critique the ways states have mishandled carbon markets, while still embracing the opening left behind by the failure of the state to deal with climate change. It's not even so much an *embrace* of that opening, in Jospe's case, but a sense, tinged with responsibility, that *someone* needs to step up and do something already. The state and the international regime should have set a price on carbon, but they haven't. Entrepreneurial projects like Nori can thus be seen as a response to abject failure by states. And while a mistrust of entrepreneurs and entrepreneurial logic is understandable in the wake of characters like Russ George, the entrepreneurs themselves might be the wrong focus of critique. Indeed, the figure of the socially conscious entrepreneur will play a vital role in the near term.

What is Jospe's long-term vision for Nori? "Our hope is to be a middleman that gets rid of the middlemen, and then gets rid of ourselves ... we can set up the framework and then walk away," Jospe explains. There's an interesting tension here between simplifying and democratizing the accounting of carbon removal, on one hand, and creating green collar jobs, on the other. This is just one instance of a deeper twenty-first century challenge: locating good jobs in an era of increasingly cheap machine labor.

Machine labor: Carbon removal in a world of advancing automation

When will artificial intelligence outperform humans? According to a survey of hundreds of machine learning experts, this could be sooner rather than later for many tasks: "translating languages (by 2024), writing high-school essays (by 2026), driving a truck (by 2027), working in retail (by 2031), writing a bestselling book (by 2049), and working as a surgeon (by 2053)."[18] These experts also believed that there was "a 50 percent chance of AI outperforming humans in all tasks in 45 years and of automating all human jobs in 120 years, with Asian respondents expecting these dates much sooner than North Americans."

Whether or not you agree with the assessment of machine learning researchers—a sample that is both knowledgeable and biased—there's a fair case to be made that machines will be doing various parts of this carbon removal work over the next century. This is both fantastic and unfortunate.

Automation could be key to the cheap scale-up of carbon removal.

Robotic labor could help to build the parts of carbon dioxide removal infrastructure that would be mass-produced, like direct air capture machines, making it a cheaper and more doable rapid transition than past energy or technology transitions would predict. There is actually a confluence of fields—robotics, the "Internet of things," and artificial intelligence—that could hold the potential to make a carbon removal scale-up cheaper—that is, if it were applied to this problem. Currently, Silicon Valley is busy with other things besides the critical plight of our planet. However, robotics is being applied, for example, to forestry (via drones), and it could be used in underwater cultivation as well as monitoring. Smart sensors could help with monitoring carbon flows and stocks. Microsoft's AI for Earth initiative, for example, has all kinds of environmental informatics applications for precision conservation, such as monitoring runoff or species tracking, which would help people understand the impacts of various carbon removal strategies.

On the other hand, broad reliance on automation—in both mechanical and knowledge jobs—could mean that a society grounded in carbon removal may not be transformative to the workforce in the ways we might hope. In fact, the trends in automation caution *against* the application of a "green jobs" justification to cheerlead carbon removal.

But carbon removal as a massive, mission-driven project would create some jobs, and taking a worker-centered lens to the endeavor would help ensure that these were decent green jobs, rather than poorly paid or hazardous ones. It's challenging to think through this, though, since our notions of work will probably be undergoing massive revision at the same time.

In one version of a post-work world, as Nick Srnicek and Alex Williams write in *Inventing the Future*, we can foresee a surplus humanity. There are more urban migrants, more circular labor, more expulsion from land and the formal economy into informal, low-paid, and irregular work.[19] "Premature deindustrialization," they fear, is leaving the urban proletariat dispossessed of their agricultural livelihoods, but without the opportunity to obtain manufacturing jobs. Under capitalism, they note, jobs have been pivotal to our social lives and identities: "Our inner lives, our social world and our built environment are organized around work and its continuation." One option for coping with the loss of jobs is to institute a universal basic income, which, they explain, is a political transformation, not only an economic one. But it matters how this basic income is instituted—implement a universal basic income too soon and too low, and it would act as a handout to companies.

Another vision sees work continuing despite the automation of labor, but in a transformed state. Tech journalist Luke Dormehl points to an "artisan economy" that favors "high-touch jobs."[20] Thomas Davenport and Julia Kirby, in a book called *Only Humans Need Apply: Winners and Losers in the Age of Smart Machines*, ask: "If you were a machine, what shortcomings would you readily admit to, and wish to have a human making up for?"[21] In what reads like a how-to manual for surviving the job apocalypse, they suggest that the job seeker ask not "What work can I do without a computer?," but "What work can't be done by a computer without me?" Their answers: pursue design and creative thinking, provide a big-picture perspective, interrogate and synthesize multiple systems and results, elicit information, persuade humans to take actions, and deal with exceptions to rules—among others. They point to the increased value of traits like empathy, humor, ethics, integrity, taste, vision, and the ability to inspire. The issue with cottage, boutique, or artisan applications of carbon removal, like one might see in some regenerative agriculture products, is the ways this comes into tension with the massive scale of the task at hand.

Besides these "abolish work" and "we'll just create new jobs" schools of thought, there's also a line of thinking that considers the effect of artificial intelligence on employment to be overblown. As sociologist Judy Wajcman observes, an oft-cited estimate that 47 percent of US jobs are vulnerable to automation within the next two decades is based upon an algorithm that predicts the susceptibility to automation of different *occupations*, rather than the *tasks* that individual jobs entail. "The popular commentators and journalists, not to mention the business consultants, seem to devour this bleak picture with a Frankensteinian relish," she notes, pointing to the "pleasure—even pride—we take in the idea that a man-made, robot-worked utopia/dystopia is on its way."[22]

It is not yet clear which of these three versions of our work future will prevail, but I find it highly likely that the social response to automation and machine intelligence will inform the shape of carbon removal jobs. There will be even more questions about why we work, what actually defines a "good" job, and what sort of labor is necessary. Carbon removal jobs could at least offer a form of labor with a clear mission and sense of social good. If fairly compensated, these would be socially valuable jobs; and in that sense, they might be "good" jobs for this century and the next.

Nature's labor: Counting on life

The work of carbon removal might be orchestrated by humans, but it hinges on the participation and enrollment of all kinds of life forces. Literature on lively nonhuman labor has been blooming in recent years. For example, anthropologist Stefan Helmreich writes about cyanobacteria, and how blue biotech companies configure microbial biodiversity as a form of accumulated labor power. He describes how the reproductive capacity of marine microorganisms can be "channeled into profit-making commodities and accumulation strategies (contrast biocapital with necrocapital: dead matter, like fossil fuel, put to unregenerative, zombie-like work)."[23] Putting the focus too heavily on the power of organisms naturalizes biotech: productivity isn't the essence of these creature's species being, Helmreich points out, and they're not just *naturally* factories or assembly lines. They only become so in certain relations. It's knowledge work, on the part of humans, and our enclosure of the organisms that make them do labor for us.

Another social scientist, Elizabeth Johnson, describes how the imaginary of planetary management enrolls nonhumans as workers, rather than "attending to multispecies entanglements with an ethics of care." The capacity for transformation of earth systems has always been more than human:

Workhorses, oxen, mules, bees, etc. have and do work alongside the living labor of humans. Like machines, these organisms facilitate the transformation of the earth's material, adjusting the parameters of space and time. With them, we have accelerated planting and harvesting, extraction, acts of war, and migration across landscapes. But other-than-human organisms are not merely 'natural resources,' the products of human innovation, or engines of extraction. They are biological entities that we have shaped and that shape us. At times their labor or lives are appropriated in ways that enhance human life.[24]

What is the right way to treat our nonhuman worker-partners? Talk to the scientists and growers working on carbon removal, and you'll see notes of affection, of commonality. Perhaps this will evolve into a code of behavior, into certain norms (or even rituals) that are passed on. Education, as we'll discuss in the next chapter, is key to the adoption of an ethics of working with other life-forms.

Both automation and the enrollment of nonhumans via genetic modification and cultivation techniques can be seen as new frontiers of accumulation. Jason W. Moore, theorist of what he terms "world-ecology," argues that we are at the end of "cheap nature," representing an epochal crisis for capitalism. The increasing necessity of carbon removal, too, indicates that we're at the end of "cheap nature"—it signals the arrival of the point where we have to pay to remove the carbon, and it's no longer cheap. One question is whether these other frontiers—these other ways of generating surplus value—create enough revenue to pay the cost of the carbon removal. Without redirection of those surpluses via a concerted political struggle, this appears unlikely. The next question is: Will those other frontiers produce enough surplus to pay for carbon removal, and universal basic income, and renewable energy, and poverty reduction, and …everything else we need and desire? This is why degrowth, and the consumption patterns of the rich world, need to be part of the conversation around an after-zero society.

The people's labor: Or, I don't want to clean up after their fucking party

"We are as gods and have to get good at it," wrote Stewart Brand, in his 2009 book *Whole Earth Discipline*.[25] This line is evidence of hubris, according to geoengineering skeptics. The line has also been quoted people by who believe the Anthropocene to be some kind of step-up-to-the-plate moment for both earth systems *and* social transformation. Brand's original passage from the 1968 *Whole Earth Catalog*, however, is seldom quoted in full:

> We are as gods and might as well get good at it. So far, remotely done power and glory—as via government, big business, formal education, church—has succeeded to the point where gross defects obscure actual gains. In response to this dilemma and to those gains a realm of intimate, personal power is developing—power of the individual to conduct his own education, find his own inspiration, shape his own environment, and share his adventure with whoever is interested.[26]

Today, indeed, much of the development of the carbon removal field is led by individuals following a path or calling. It's quite resonant with the 1960s wish that individuals have the power to shape their own environment. Yet this wish is in tension with the demands of the task at hand: the removal of gigatons of carbon. Are these going to be artisan jobs, performed by practitioners of a craft—as opposed to "remotely done power and glory"? Or does the scalability requirement demand remote control and configuration, orchestration? In the end, there will have to be serious engagement by the state. With all this talk about scalability, where is the room for the non-scalable? The bespoke? The self-determined? Or is the self-determination of the entrepreneur, the making of one's own destiny, only a mirage?

I will close this chapter with three brief contentions—really just seeds for further discussion. First, the people who will be doing the work of carbon removal should be defining the field, and what sort of jobs these are. Workers should be able to set the terms of their labor, and following from that, make a decent living providing the service of carbon removal. We can start by bolstering unions and experimenting with collectively owned factories and farms. Second, and at the same time, we need open-source technologies that can be adapted by people in various contexts around the world, because the terrain for innovation and the capacity to implement carbon removal is egregiously uneven. Third, creating good jobs in carbon removal will need to include attention to race, gender, and inclusion.

I recently attended a lecture at a modern art museum—the works of art were luminous orb-like freezers, displaying slabs of meat, skeletons, and plastic artifacts, meant to evoke Anthropocene strata. The scholar lecturing on the exhibit called for better housekeeping in the Anthropocene. I knew what he meant, sure: the mess needs to be cleaned up, the earth-home cared for. But the word "housekeeping" brings up the fact that housekeeping is gendered and racialized work. In the lecture's suggestion of an Anthropocene housekeeping, there was no mention of race or gender; nor the fact that the makers and beneficiaries of the Anthropocene have been primarily white men. These men made a 2-trillion-ton mess with fossil fuels, profited handsomely from it, and then are going to go off and smoke and drink tea and whiskey, while we take care of the dishes. (Not All Men, sure; and some women somewhere benefited—washing machines! suburban landscape through which to drive to soccer matches!—but the basic contours are accurate.) Except it's complicated, because the rich white men also own the industrial means and control the technologies that we would need to clean it up. So: (a) either the men get a double-profit, or (b) they simply won't get around to cleaning it up, while climate suffering falls disproportionately on people of color and women. The nascent carbon removal technologies are by and large controlled by white men: both the basic materials used in the carbon removal methods discussed in the previous chapters, and the technologies described in this chapter that modify carbon removal's implementation. The remedy for this situation can start in many places. One is at the beginning: in childhood, with socialization and education.

7

Learning

...the kind of thinking we are, at last, beginning to do about how to change the goals of human domination and unlimited growth to those of human adaptability and long-term survival is a shift from yang to yin, and so involves acceptance of impermanence and imperfection, a patience with uncertainty and the makeshift, a friendship with water, darkness, and the earth.

—Ursula K. LeGuin

Pacific Grove, California, January, 13°C / 55°F

Outside, it's misty and gray, spitting salt spray and rain. Beyond the dunes, Pacific breakers slam into the land. On this side of the sheltering dunes, there's thin soil, slick rock, lichen, sick pines oozing resin.

I'm inside a solid-stone lodge, warming myself beside a crackling fire. A woman in sturdy boots pokes the logs, sending lazy sparks aloft. Another woman, bundled in a scarf, is knitting soft gray fiber with a circular needle. The rows of stackable conference chairs upon which we sit, the quiet podium with its waiting PowerPoint, and the sturdy beams evoke a collision between a nineteenth-century living room, a hotel meeting salon, and a church.

Asilomar is a conference center on the Monterey Peninsula. In science policy circles, it is known as the site of a 1975 meeting in which scientists came together to discuss the risks of recombinant DNA research, agreeing on guidelines that formed the basis of US regulation of the field. This researcher-led model of governance was the inspiration for a 2010 meeting of geoengineering researchers at Asilomar, which attempted to deal with standards for climate engineering research and assess the risks of experiments. I've come to Asilomar for quite a different reason, though: a workshop exploring regenerative agriculture. In fact, it's quite the opposite of top-down, expert-led governance of climate intervention. It's about intervening quite literally from the ground up.

The proceedings begin with an invitation to get grounded. We place our feet on the floor and meditate on the question: "What do you want to regenerate?" Sharing time: the woman next to me says she's interested in understanding the flows in nature. Wow, I think, what doesn't need to be regenerated? The plants in my backyard, after just one rain event in the past six months, are looking pretty sad. Not to mention the blackened, fire-blasted hillsides of Highway 101 that I drove past on the way here. My cells could probably do with regeneration. The country probably needs regeneration. I'm overthinking it: I say something about community, generic but true. Then, the energetic speaker, soil scientist Ray Archuleta, begins.

"Good morning! This is an exciting new movement!" He's talking about the soil health movement—and, in particular, the move from sustainability toward regeneration. "If somebody says, how's your marriage, and I say … Oh, it's sustainable … what do you think?" Everyone laughs. "Ladies and gentlemen, I'm blessed that my marriage is regenerative." Archuleta then talks about the marriage between us and the land: "It starts with people of integrity." For him, the revolution started with a realization of "failure." As a government employee working in soil conservation, he began to see that what they were doing wasn't working. "Reductionist conservation," such as filter strips, riparian forest buffers, windbreaks, and grass waterways, wasn't getting the job done. "We have made yield our god, but look what's happened to our communities!" You drive through the skeletal remains of rural America, he says, pacing, and see the death of community, along with that of the fossil agriculture powered by oil from Canada's tar sands, "Why are America's farmers killing themselves in record numbers?" Ray asks. "The stress, the pressure." The whole system is broken.

"We have all been products of Western thought processes since Aristotle," Ray declares, explaining how reductionist thought has been ingrained through years of filters. Universities, governments, parents, grandparents—all living with the wrong paradigm. "A good paradigm will smash yours," Ray says; it will smash through your filters.

Ray's paradigm is the chaotic, elegant, and beautiful paradigm of soil health. "We are all dirt on legs." The word human is from the Latin *humus*. "Folks, it's us. A living substrate." We are all interconnected. We must allow the soil to speak to us: "It will tell you, if you listen." He refers to Job 12:7: "But ask the animals, and they will teach you, or the birds in the sky, and they will tell you; or speak to the earth, and it will teach you, or let the fish in the sea inform you."

Then, Ray commences his demonstration. For audience members who may not grasp how to "listen" to soil, it's a fairly simple visual display known as the "slake aggregate stability test." On stage, there are five large cylinders filled with water. Five people are chosen from the audience. The volunteers gently drop different clods of soil into the columns of water: one clod from North Dakota, one from Missouri, and one that was cultivated using no-till agriculture. Which holds its integrity and structure? The visual evidence illustrates how the soil clod with no-till holds its integrity. "Wow," someone whispers behind me.

"The tillage is so brutal," Ray explains. Tillage, indeed, is an intrusive tool. There's another demonstration, of rain. "Ladies and gentlemen, the water does not infiltrate," he announces. The tilled soil can't absorb the water. "We do not have a runoff problem; we have an infiltration problem."

"Look in the mirror: we're the problem. We're disconnected."

What is the difference between the functioning soils and the leaky ones? "The understanding of the producers," Ray emphasizes. Regeneration begins in the heart and the mind. This is echoed by the next speaker, a North Dakota farmer named Gabe Brown, who practices regenerative agriculture on his ranch. "The revolution starts with a few," Brown says, explaining that our carbon is out of place.

"If you want to make major changes, change the way you see things." Brown says he can tell right away when someone is ready to learn—that's the way it is with regenerative agriculture—and quotes the spiritual adage, "When the student is ready, the teacher will appear."

"The real product is the mind of the farmer," I kept hearing throughout the day. This idea of changing one's way of seeing was primary. It wasn't only about learning, but unlearning.

How do we learn, not just the skills, but the mindset that would be needed to realize a carbon removal society? What would it mean to culturally internalize the aim of carbon removal—to have it as part of the fabric of everyday life—and how do we get there? These questions might seem eccentric to some readers. After all, much of carbon removal looks like familiar day-to-day work. As Klaus Lackner says of direct air capture, "I think the people that make this work are not all that different than people who make windmills work, who keep solar panels running, who make solar thermal plants work. And so there are a lot of jobs that are parallel to that. But of course, in the details, they are different, right?" These are similar activities: electric work, plumbing, creating automated devices, writing software, and fixing it all. "To a large extent, there's a manufacturing industry behind it. But, I don't think it takes all that much a different set of skills than building cars or building washing machines," Lackner says. From a purely technical standpoint, you could imagine a carbon removal society that does the same kinds of activities as today, with an economic fabric that simply replaces carbon taken from the ground by orienting activity toward mining it from the air. That substitute for real transformation is, of course, exactly what many social justice advocates fear. And in my view, not only would replacing one commodity for another be a missed opportunity—it would likely fail to reach net-negative levels.

Right now, fossil fuels are embedded in the structure of our political systems, economies, and everyday lives. If a society patterned around fossil fuel extraction and greenhouse gas emissions has a specific way of working, as some thinkers postulate, it figures that a society patterned around drawing down carbon would also have specific ways of looking, feeling, and working. As political theorist Timothy Mitchell points out, the leading industrialized countries are oil states; thus, without energy from oil, their current forms of political and economic life would not exist—food, travel, housing, consumer goods, and more.[1] In Mitchell's analysis, oil produces both specific forms of democracies and the economy itself. It's not just fossil fuel extraction that shapes the patterns of our lives, though. As Carbon180 executive director Noah Deich pointed out in our conversation, there's a wider "greenhouse gas emission paradigm, where it's not just fossil fuels but clearing forests, farming in ways that release carbon from soils, et cetera."[2]

In many ways, a world patterned around carbon removal would be similar to one that's committed itself to deep decarbonization and extreme mitigation. Under one view, it would essentially *be* that world, taken a step further. In another view, it might be qualitatively different, because regeneration, removal, restoration, and so forth bring a different narrative than mitigation, and perhaps a different politics. How? First, it may be easier to build a broader coalition around regeneration. Second, the goal is more drastic. As Deich notes, there is currently a mass movement toward decarbonization. "We're just not aiming for the end zone yet. We're aiming for the twenty-yard line, and that doesn't work." It might seem nonintuitive that a more drastic goal has a better chance of success, but it is possible that it's more galvanizing.

It would be extraordinarily speculative to imagine the contours of carbon removal cultures (and they would certainly be plural). We can begin to think about it by considering education. Like carbon removal, education is a long game.

Right now, both the primary and university education systems are broken in several ways—and in many areas of the world, certain groups don't even have access to education. Dozens of other books analyze and diagnose the ills in these varied educational contexts, from early childhood to the university. For the purposes of this discussion, I note six major challenges to the transformation of education. One is disciplinary siloing. Carbon removal, like many other socio-environmental challenges, requires an ability to understand across disciplines. Second, formal education at every level needs to be decolonized. A third challenge is the rote learning still present in primary school education, which inhibits creative problem solving and the innate capacity for inquiry. Fourth, higher education enrolls many students in debt, curtailing their futures and limiting their choices, at least in the US context. Fifth, teachers are undervalued, which is an obstacle to quality education in general. Sixth, as jobs become automated or obsolete, people will need reskilling throughout the life course in order to shift careers. One could add to this list for a very long time.

Yet despite all the ways in which this system is broken, people maintain faith in education as a remedy for both social and environmental malaise. What other choice do we have? Recently, I took part in a discussion with Silicon Valley thought-leaders about the threats posed by increasing inequality. It was chaired by a CEO of a major tech company, who expressed genuine concern about the trends in income disparity. The conversation, though, kept coming back to education as if it were some kind of salve or universal remedy—if we just educate people right, it will fix inequality! Education is a quintessential liberal panacea for deeper structural problems. Thus, I do not wish to insinuate that it is some kind of cure that obviates political change. Rather, education is tied up in the political change we need. It is therefore hard to imagine climate repair taking place without deep changes in our education. Both the content and the methods of instruction need to be shifted, throughout the life course, to create educational experiences that can meet twenty-first century socio-environmental challenges.

Let's continue to imagine a world that has embarked upon carbon removal for climate restoration, and consider what they had to learn and unlearn to get there. In the following thought-experiment, I identify ten key capacities to build during early formal education, which might make realization of a carbon removal society more possible:

1. *Critical design skills, including visualization of the invisible*. This encompasses a basic concept: we need to be able to imagine and consider something that we can't see, carbon. But the same could go

for seeing microbial life, or for seeing people in other places that are impacted by our actions, such as the workers producing our goods. Here I refer not only to the skills of being a designer—of data visualization and user interface design—but the skills of being a user and critic of others' designs. For example, imagine devices that track carbon flows—this could be a positive, fun game, or a nightmare of calculation, depending on the design. Artifacts like the Styrofoam cup, the internal combustion engine, and others become inelegant from a design perspective (assuming, of course, a universal design aesthetic that is focused around environmental sustainability or regeneration). If the culture at large were infused with a design sensibility, it could alleviate part of the problem—demand for wasteful, polluting things and landscapes—and moved in the direction of carbon removal.

2. *Empathy—cross-cultural and multispecies.* This competency comes with visualization of the invisible, but transcends it, in that it also extends to care for others—including those displaced in space or time. At present, preschoolers seem to be the only ones coached in empathy, but this faculty could be practiced throughout education.

3. *Decolonial practice.* Basic knowledge of history and geography are a foundation for this. Understanding fundamental processes of colonialism and exploitation will help students grasp the limitations to carbon removal, as well as the equity issues involved. This understanding will be key to seeing climate restoration actually happen. For understanding history opens space and insight for other possible futures. Decolonial practice is about more than just content knowledge, though; it's about the process of recognizing hegemony and domination, and acting on those understandings.

4. *Experiential knowledge of the natural world.* This is necessary for having a true relationship with the outside world.

5. *Numeracy and scale.* In the context of a shifting climate, it will be essential for the new generation to understand the scale of the climate change problem and the methods proposed for addressing it. This including the capacity to work with big numbers as well as orders of magnitude, for gigatons, millions of hectares, and trillions of dollars are still nonintuitive to most people. This probably has to

do with the way we teach math in primary school, where the focus is on operations rather than development of mathematical intuitions.

6. Critical algorithmic literacy. At root, critical algorithmic literacy has to do with understanding models for predicting the future— and the decisions that are based upon those models. People need to learn the limitations of these algorithms, how to intuit when they might be wrong, and how to improve upon them. This isn't just a technocratic, expert-level skill for programmers; it will need to be a fundamental part of life in a society that is, for better or worse, partially governed by algorithms.

7. Interdisciplinary systems thinking. A society embarking on something as complex as restoring the climate will have to comprehend myriad interacting human and natural systems, from carbon and nutrient cycles to hydrology to microbes to economies.

8. Dialogue. Carbon removal will never be realized unless people build coalitions to do it, which means dissolving ideological boundaries and political or identity-based factions, and working together on an agreed-upon goal. An educational system and scholarly discourse designed solely around constructing arguments and critique, rather than listening and collaborating, is doomed to fail at a task such as this.

9. Imagination. While students are increasingly taught the kind of creativity and imagination that is useful in entrepreneurship, a related capacity is the ability to imagine large-scale or long-term changes in systems. What does it take to imagine a world that's quite different from the one we live in now? What activities or practices enable and strengthen it?

10. Emotional self-knowledge. There is tremendous emotional content and context around climate change: fear, loss, guilt, vulnerability, love, and longing, to name a few. Grasping this, and being able to connect with and articulate it, will probably be necessary to mobilize around the scale of action called for.

Cultivation of these capacities would constitute an education not in methods of control, management, or domination of nature, but in how to work with it. Meanwhile, the STEM engine keeps chugging; people still view learning to code as a ticket to the next income tier, even as inequality rises and computers get better at programming themselves. The trends for some of these competencies, such as critical algorithmic literacy, design, or data visualization, look positive, and they are perhaps inevitable as education transforms. Others, such as empathy, dialogue, and experience with the natural world, could easily be left out.

8

Co-opting

Baku, Azerbaijan, 2010, pleasant weather,
with oil prices over $100 a barrel

was strolling along the edge of the Caspian Sea when I happened upon the setup for Oil Worker Appreciation Day. Along
the trimmed lawns of a quiet park, wall-sized photos of oil infrastructure had been erected—gleaming refineries, offshore rigs,
compressors. National patriarch Heydar Aliyev was photoshopped
into some of them, looking inspired or determined in his tuxedo,
with the blue-red-green of the Azeri flag rippling behind him.
Breaking up the infrastructure panoramas were images of seedlings
being replanted in depleted oil fields, scenes from new hospitals, and
a map of the Baku-Tbilisi-Ceyhan pipeline. The State Oil Company
of Azerbaijan Republic (SOCAR) had also constructed stages for
evening concerts in the park. The bands would play against back-
drops of oil rigs and tambourines. I tried to picture people dancing
into the night, courtesy of SOCAR.

In some places, oil connotes regeneration. Shining blue-glass
skyscrapers were being sculpted in the shape of flames. Versace and
other fashion houses were popping up with silent boutiques, and the
streets were jammed end to end with Mercedes SUVs. Resplendent
fountains and gardens appeared in the new plazas; more were under
construction.

Traveling out beyond the capital, to the oil fields of the Absheron
Peninsula, I walked through a different picture: people living among

unfettered extraction. Baku, like many capital cities, is an anomaly. Out near the oil fields, children played in the streets, indifferent to the puddles of pitch-black oil seeps. Women strung up laundry to dry between pieces of decaying oil infrastructure, red-and-white striped towels fluttering among empty cylindrical tanks. Rusting pipelines, no longer connecting anything with anything, crisscrossed bare earth or shrubs. A newer set of pipelines, mustard yellow, connected the pumps that were still sucking. Everything was drenched in the smell of naphtha. In the midst of it all, under the flat, dusty-blue sky, women carried home loaves of round bread.

Azerbaijan relies on energy for 70 percent of its income; oil makes up 95 percent of its exports. Now, with oil prices half of what they were a few years ago, the country faces the task of developing non-oil sectors. Yet oil is so embedded that the fifty-manat banknote features a diagram of the benzene molecule, C_6H_6: six carbon atoms, six hydrogen atoms, arranged like an esoteric star.

A key question behind decarbonization is this: How does one get companies to walk away from their assets? Or: How does one get countries to walk away from assets they are counting on? For three-quarters of the oil extraction is done not by international companies like ExxonMobil and Shell, but by national oil companies like SOCAR. (The latter are actually entangled with private companies—BP owns a 20 percent stake in Russian firm Rosneft, for example.) Oil-producer governments still capture a large part of revenues—on average, 70 percent of net revenues, from 40 percent in the United States to 95 percent in Iran.[1] Geographers Gavin Bridge and Phillippe Le Billon break down the oil value chain: with oil at one hundred dollars a barrel, 20 percent of that goes to cover costs, 33 percent is gained by producer governments, consumer governments earn 40 percent, and companies get 7 percent.[2] Note also that privately held companies aren't simply private; investment by pension funds in fossil fuel companies means citizens are caught up in their fates.

If oil prices were back up to one hundred dollars a barrel, the 1.7 trillion barrels of oil in reserves add up to $170 trillion of unburnable carbon—two years of global GDP.[3] The infrastructure at stake is also worth tens of trillions. That is a lot of revenue, not just for companies, but for nation-states like Azerbaijan, to turn away from. In many cases, citizens aren't profiting from these revenues, though in some areas they are. All this entanglement means that the dilemma of what to do with these fossil fuel entities isn't a simple one of fighting against a few corporations we don't like. We need to understand it as a social question, not just a business or economics or political question. Despite the stirrings of discussion about a fossil fuel "exit" or "phaseout," this question still needs to be pushed into the mainstream. So far, the literature on "de-risking" pathways for companies is very corporate, originating from think tanks and research institutions, and aimed at fossil fuel companies. It is not a social question—despite how socially embedded these fuels are.

The entanglement of states, citizen investors, and fossil fuel producers also means that if the narrative flips and the turn away from fossil fuels truly happens, the public could suddenly be holding a lot of debt and liabilities. If we are serious about getting off of fossil fuels, we may be headed for a bailout that makes the industry bailout following the 2008 financial crisis look like crumbs. For example, consider Peabody Energy—responsible for 1.16 percent of global greenhouse emissions, and number sixteen of the 100 top producers of greenhouse gas emissions according to a list compiled by the non-profit CDP.[4] Its predecessor, Peabody Coal, went bankrupt in 2016, when coal prices dropped and they were stuck with $10 billion in debt—and their executive walked away with significant compensation. Then, the reformulated Peabody Energy "exited bankruptcy" in 2017, back in action. Meanwhile, Peabody was one of thirty-seven fossil fuel companies being sued by municipalities in California for damages due to climate change. However, a judge ruled that they weren't responsible for climate impacts incurred before their 2016 bankruptcy; they got a clean slate. It's not hard to imagine this maneuver becoming part of the playbook. Indeed, immunity from climate change lawsuits for fossil fuel companies is shaping up to be a part of any compromise legislation on climate change.

There's another reason what to do with fossil fuel entities is a complex social question, which is that these revenues are sometimes used to fund mitigation and adaptation to climate change. For example, the state of Louisiana is rapidly losing land to coastal erosion. The state's master plan for coping comes with a $50 billion price tag, and, as journalists Kevin Sack and John Schwartz report in a 2018 article in the *New York Times* and New Orleans's *Times-Picayune*, "the only dependable financing model has been catastrophe": the 2010 Deepwater Horizon oil spill.[5] Offshore leases were also budgeted in at $176 million a year, to help pay for adaptation, but those revenues have fallen with oil prices. That $50 billion is twice the state's annual budget, and a federal bailout seems unlikely: there will be competition from other coastal areas, like South Florida and metropolitan New York. Oil and gas, Sack and Schwartz report, are the only industries flush enough to fund some restoration, and despite some politicans' reluctance to hold a main source of local income accountable, Louisianans "seem newly receptive to holding the industry accountable for the consequences of its activities." In a *Times-Picayune* poll, 72 percent agreed that industry should help pay, and another 18 percent said that industry alone should bear the cost. Only half of individuals were willing to pay higher taxes for coastal restoration. But they already do: state taxpayers spent $588 million to repair oil-and-gas-related damage along the Louisiana coast, which perversely goes to benefit those very industries, as they own or lease much of the coast.

Examples like this illustrate three things. First, the cost of these climate-related, slow-onset disasters is staggering, and it will soon be a ballooning public burden. Second, people *do* support making the industry that created the problems pay, to some extent. Third, despite this idea that polluters should pay, the risk seems high that it is taxpayers who will end up bailing out the fossil fuel industry—perhaps even in ways that serve these companies. They are not going to go gently into that good night.

This social question of what to do with fossil fuel producers intersects carbon removal on the deepest levels. Carbon capture and sequestration could be a way to allow fossil fuel companies to slip into something different. Bridge and Le Billon, in their study of the political economy of oil, write that for oil producers, "serious efforts to stem the accumulation of atmospheric carbon raise the interesting prospect of them becoming stewards of underground carbon stocks rather than extractors of oil." The oil production network could become a carbon conveyer, they write, and a new class of end user could be created: those actors who own or control carbon sinks.[6]

Consider the remarks made by Charles McConnell, who was head of the US Department of Energy Office of Fossil Fuels during the Obama administration, at a December 2018 oil industry conference in Midland, Texas.

> What's the Paris Accord and what does it really do? It gets you about 0.4 percent of what's absolutely necessary to achieve two degrees by 2100. Fundamentally, what I am saying is that it doesn't do anything. It would delay getting to four to five degrees by 2100 by about four years. And it's trillions of dollars. It doesn't talk about technology; it doesn't really talk about CCUS [carbon capture, utilization, and storage].[7]

Here, the narrative maneuver is that the Paris Agreement is weak, unambitious, ineffective. There is a fair argument there. But then McConnell goes on to describe fifty to one hundred years of oil production from enhanced oil recovery, a technique of getting more oil out of depleted oil wells, which is supposed to be a bridge to some other future. What he closes with is also interesting: a warning that both national and international oil companies are pointing their strategy toward CCS with EOR and eventual decarbonization: "These companies are planning for a decarbonized future, however that gets envisioned. And the smart ones are the ones who are out in front. They are going to create their own world and they are not going to be victims to someone else's."

We've been hearing about the need to adopt CCS for years: Is this time different? It might be. The oil industry is moving into a defensive crouch. As the secretary general of OPEC said at the 2019 World Economic Forum in Davos, regarding climate change and pressure from investors: "Our industry is literally under siege and the future of oil is at stake."[8] The head of state-run Saudi Aramco stated, "We need to boost efficiency or get rid of CO_2 by technology." Even though the world failed to invest in CCS for decades, the moment of this narrative shift—where fossil fuel becomes something dangerous to invest in—might place pressure on oil companies to take the lead on carbon removal, if only to preserve investor confidence (whether or not they actually follow through on doing it).

What if fossil fuel companies transform into carbon management companies? How would that shape the future of carbon removal?

In a worst-case scenario, oil companies will adopt the logic and argument of negative emissions in order to keep producing oil with enhanced oil recovery, employing a discourse of "carbon management."

To see a more specific, less speculative example of how policy for CCS is in thrall to vested interests, take the response to the FUTURE Act in the United States. This legislation was passed in February 2018 as part of a federal budget bill. It reforms "Section 45Q," the US tax credit for carbon capture and storage. Previously, CCS projects received ten dollars per ton of carbon captured and used for enhanced oil recovery, and twenty dollars per ton securely stored. The revision increases these sums to thirty-five and fifty dollars, respectively, and eliminates a volumetric cap of 75 million tons of CO_2. Following the successful changes to the credit in the new US budget, a new coalition was launched: the Carbon Capture Coalition, which is the rebranded National Enhanced Oil Recovery Initiative. The group boasts forty-eight members, including companies like Archer Daniels Midland and Mitsubishi Heavy Industries America, fossil fuel companies like Peabody Energy and Shell, think tanks like the Bipartisan Policy Center, and unions like the AFL-CIO, the National Farmers Union, and the International Brotherhood of Electrical Workers. But it lost a member—the Natural Resources Defense Council, a green group that doesn't support incentives for enhanced oil recovery.

There are some reasons to celebrate the legislation: it was a rare bipartisan effort, and it is a performance-based tax credit, meaning that something measurable has to actually happen for it to be claimed (though these companies don't pay a whole lot of taxes to begin with). But the whole idea of having tax credits for EOR illustrates the challenge: the polluters are first in line to benefit from carbon capture. A letter signed by organizations 350.org, Greenpeace USA, Clean Water Action, Friends of the Earth International, and others called the act a "handout to oil companies." They also pointed out that EOR negatively affects people of color and environmental justice communities, who disproportionately live near oil fields.

If there's no progressive vision about how to use CCS, including a clear set of demands about how we want to use this technology, the oil companies can essentially take us hostage. For if we don't need them for fossil fuel extraction, we will need them for removal services, since they have the very technical capacity that's needed to inject CO_2 underground: the drilling and seismic expertise, the work in offshore environments, and so forth. They are the ones that have developed and pioneered this technology—with heavy government subsidies and investment, of course. It'll be: *Everyone knows climate change is bad; the international agreements aren't strong enough and aren't achieving enough; this carbon needs to be removed; and we're the only ones with the expertise, know-how, and get-things-done spirit to accomplish it.* It is better to meet these prospects head on, in an anticipatory way, rather than pretend we can wave a magic wand and ban all fossil fuels, or vigorously oppose all forms of industrial carbon capture. What's needed is a specific set of terms for CCS—to start, I would personally throw out "no CCS with coal," but it's clearly a collective conversation. We can see the fossil fuel industry organizing to take charge of the narrative shift towards acceptance of carbon removal. There is already bipartisan legislation introduced in the US Senate to incentivize large-scale carbon removal—but it's called The Enhancing Fossil Fuel Energy Carbon Technology (EFFECT) Act, which has been introduced but not turned into law yet. It defines "net-negative carbon dioxide emissions technology" as technologies that co-fire coal and biomass and places ensuring the continued use of coal as a goal—this is the bargain that Democrats are asked to sign onto to get a cross-cutting carbon removal program funded. We need to organize an alternative narrative about how we think carbon removal should proceed, before the line that "there is no alternative" to the vision set out by the fossil fuel companies rears its ugly head. The longer we wait, the more entrenched their vision will be.

Retiring entrenched powers

Fossil fuel companies aren't the only actors that might co-opt carbon removal efforts. Once there is a price on carbon, and a value assigned to it, then the jostling will begin in earnest. First, there's industry, responsible for about 20 percent of greenhouse gas emissions, with heavy contributions from cement, steel, and fertilizer, in particular. This sector has a history of lobbying against a price on carbon, yet it may eventually embrace carbon removal as a way of continued existence, getting exemptions for "residual emissions." Next, there's the farm lobby. Journalist Stephanie Anderson's book *One Size Fits None: A Farm Girl's Search for Regenerative Agriculture* explains the political work to be done to move toward regenerative agriculture: the farm bill needs to fund research in regenerative agriculture, the land-grant universities need to be redirected and freed from corporate influence, the Cooperative Extension Service needs to be retooled, and subsidies for farmers need to be changed.[9]

Then there are the indirect threats to scale-up of carbon removal. At first glance, defense contractors don't appear relevant to this topic. But in the United States, we fund the military to the tune of $500–700 billion each year, a significant proportion of which goes to contractors, who in turn lobby for more. Can we afford to keep paying off these behemoths, and still have enough public funds to instantiate a massive decarbonization and carbon removal program? Tech companies are another group of actors that could confound a socially just version of carbon removal, though this, too, is a very indirect link. My concern is that tech companies could put forth a set of proprietary tools for managing carbon before we have time to invent messier but more democratic alternatives—thus providing a captive platform upon which carbon transactions take place. One agtech company, Indigo Ag, has launched a "Terraton Initiative" to store a trillion tons of carbon in soils—orchestrated on its platform.

Once there's a price on carbon, there's a rationale for more tech companies to engage, and they have far more capacity to do some of the data management that would be involved. Companies like Google, for instance, already have an incumbent's advantage when it comes to geospatial data. Tech money has poured into precision agriculture, and it seems a reasonable step from precision agriculture to carbon management. In their book *Climate Leviathan*, social theorists Joel Wainwright and Geoff Mann describe a planetary sovereign, "a regularly authority armed with democratic legitimacy binding technical authority on scientific issues, and a panopticon-like capacity to monitor the vital elements of our emerging world: fresh water, carbon emissions, climate refugees, and so on."[10] If their vision comes to pass, it will be tech companies who deliver those capacities. Who writes the code, and how they do it, matters to the outcomes.

It's exhausting to think about all these entrenched power interests; clearly, the path toward any socially just form of carbon removal is fraught. But thinking critically about the power of all these sectors is important, because it illustrates how significant levels of carbon removal will require more than just business as usual. They will require taking back our democracy and the levers of power from Big Agra, Big Oil, beltway bandits, and the like.

Putting carbon removal into the Green New Deal

The good news is that all kinds of people are already fighting on all of these fronts. In Naomi Klein's book *This Changes Everything*, she describes "Blockadia": interconnected pockets of resistance to extreme fossil fuel extraction. "Blockadia is not a specific location on a map but rather a roving transitional conflict zone that is cropping up with increasing frequency and intensity wherever extractive projects are attempting to dig and drill, whether for open-pit mines, or gas fracking, or tar sands oil pipelines." Blockadia is a critical emergence, and it demonstrates a valuable set of tactics. What Blockadia does, write environmental scholars Marco Armiero and

Massimo de Angelis, is shatter the universalism of the Anthropocene narrative and bring attention to race, class, and gender (as well as settler colonialism).[11] And yet, many of the victories thus far—in terms of extraction plans being shuttered or squashed—likely have to do with low oil prices. When prices rise again, we can expect a more voracious extractive advance upon areas that aren't currently economical to develop, including more CO_2-EOR projects—that is, unless public opinion and legal pressures transform the picture dramatically. I personally believe there is a gradual movement in public opinion taking place: there will be more lawsuits, more divestment, more shaming, and a shifting tide. We're not quite there, but by the time this book is printed, it will be even closer.

We need Blockadia—it's helped transform the scene to the point where OPEC appears to be under siege. But for handling carbon removal, we also need a resistance that is more than reactive. We have to move from reflexive opposition of new technologies toward shaping them in line with our demands and alternative visions. To realize a just form of carbon removal, we need to challenge fossil fuel extraction *differently*—to take the question of what will happen to these companies head-on, and transform them ourselves. What does that look like: Nationalization? Negotiation of allowable "residual emissions" for the sectors *we* decide are key? Public subsidies for carbon removal infrastructure; public orchestration of certificates of obligation for carbon disposal? Or more radically: Requiring emitters with historical responsibility to pay for carbon removal? These processes may not be as photogenic or emotionally legible as Blockadia, though with some creativity we could make them so. In general, environmentalism has focused on consumption, local alternatives, and to some extent, companies—but fossil fuels and industry are more structurally integrated than these approaches alone can address. We need to discuss nationalization, as well as strategic diversification into other areas. While companies haven't done so well with diversification into alternative energy so far, in a low-carbon scenario, it may be the best option for them. Or, why not combine those options: nationalize these companies, and turn them into combined energy / carbon removal companies? What's clear is that there needs to be a broader social discussion that (1) recognizes how entwined these companies are with the state; and (2) anticipates bailout actions or clean-slate bankruptcies in a forward-thinking, offensive way, rather than reacting to them as they unfold. Collective decision making on demands for the arrangements of drawdown and regeneration will be critical if we are to see carbon actually get cleaned up.

Nina Power, a cultural critic and philosopher, points out that to make a demand from somebody is often to accept the broad outlines of the existing situation: "To demand something—better working conditions, political representation, compensation—is at the same time often to recognize the framework and the institutions that could (but most often will not) acquiesce to that demand: employers, the government, the state."[12] The process of articulating demands is worth doing, because it helps us understand which institutions won't acquiesce to the demands, and think about how to deal with them.

Of whom do we demand carbon removal? The state, first and foremost. We can also issue demands related to carbon removal toward politicians (on various scales), investors, and companies with relevant expertise. Specific demands could include near-term policy actions, as well as public investment. Changing the subsidies for fossil fuels is what people point out as the first and most obvious step to decarbonization: the world currently subsidizes fossil fuels at $500 billion per year, or $15 per ton of carbon dioxide emissions. We should be paying for the damages instead of subsidizing what's driving them. We should also be continuing the pressure to divest from fossil fuels, while pointing out social investment opportunities in carbon removal. Right now, there's an awareness problem: investors aren't aware that carbon budgets exist, or what they mean for high-emission companies.[13] Therefore, we need to create comprehensible accounts of the risks to investors—and many are working on this—while also suggesting carbon removal among a range of long-term things to invest in. In California, Assembly Bill 1550, building on Senate Bill 535, requires that 25 percent of funds from the state's cap-and-trade program go to projects within disadvantaged communities, and another 10 percent go to benefit low-income households and communities. These types of legislative action, driven by many environmental justice advocacy groups, can provide ideas for how carbon removal funding could proceed. Actions like these would be initial steps toward making the benefits from carbon capture programs accrue to people who have suffered from environmental injustice, and toward alleviation of inequality during the transition to a carbon-negative society. In short, we need anticipatory engagement that doesn't simply anticipate carbon removal as an emerging technology in need of R&D or investment; we need anticipatory engagement that can also anticipate its co-option. In this regard, climate restoration requires an expressly political engagement.

The risks that attempts at carbon removal will fail are grim, as a chorus of analysts warning about "betting on negative emissions" have noted. And those risks are only compounded by another prospect on the horizon: solar geoengineering.

Part IV: Buying Time

9

Programming

Interface evolves toward transparency. The one you have to devote the least conscious effort to survives, prospers ... The real-deal cyborg will be deeper and more subtle and exist increasingly at the particle level, in a humanity where unaugmented reality will eventually be a hypothetical construct, something we can only try, with great difficulty, to imagine—as we might try, today, to imagine a world without electronic media.

—William Gibson

Coral reefs smell of rotting flesh as they bleach. The riot of colors—yellow, violet, cerulean—becomes ghostly white as their flesh turns translucent and falls off, leaving the skeletons underneath fuzzy with cobweb-like algae.

Corals exist in relationship. The coral animals live in symbiosis with a type of algae. During the day, the algae photosynthesize. During the night, the corals open their mouths and catch passing food. Just one degree Celsius of ocean warming can break down this coral–algae relationship, for when it is placed under stress, the algae leaves. Without the algae, the corals "bleach." After repeated or prolonged bleaches, corals starve and become diseased. Eventually, they reach an unrecoverable dead state.

Australia's Great Barrier Reef, actually a 2,300-kilometer system made up of nearly 3,000 separate reefs, has suffered severe bleaching in the past few years. Daniel Harrison, an Australian oceanographer, is looking at what might be done to buy more time for the

Great Barrier Reef. "When the reefs got bleached back to back so badly, two years in a row, we kind of just formed a little informal working group, and were like, 'You know, if we can send people to the moon, and to Mars maybe soon, then surely we can stop the reef from bleaching, you know? From just overheating.'" The situation is getting dire. "There might be as little as 25 percent of coral cover left from pre-anthropogenic times," he tells me. "We don't really know, because nobody started surveying before 1985 … It's incredible, isn't it? I mean, you've got less than 1 percent of the ocean in coral reefs, and 25 percent all marine life. And it's not just the Great Barrier Reef that's in dire trouble, obviously. You know, we're looking at losing all of that really quite quickly, in evolutionary terms. Quite quickly, in human lifetime terms."

The Australian working group formed teams to look at different ideas that could help the reef stay alive. Their investigations showed that most of their exploratory, out-of-the-box ideas wouldn't scale too well. For example, they wondered: Since the ocean is full of cooler water at deeper depths, could we just pump up some of that in order to cool of the reef? Harrison explains: "You might be able to do something to protect small areas of the reef. Or to maybe protect important coral larvae source reef, and that sort of thing. But none of the other ideas—I mean, it's just infeasible to move up enough cool water to kind of cool the whole reef." After considering different options, the researchers honed in on the idea of marine cloud brightening—a form of solar geoengineering—as something worthy of further study. Brighter, more reflective clouds could cool the area. If small salt particles were sprayed into the air, tiny water droplets could condense around them, and these micro-droplets would make the clouds brighter. Harrison is doing modeling research to better understand the feasibility of this idea. The first stage of the modeling his team has done indicates that it might be possible to cool the water by $0.5°$ to $1°C$.

Another research effort, the Marine Cloud Brightening Project, thinks that this could be a scalable approach with some promise for reefs. I talked with Kelly Wanser, the executive director of the non-profit organization SilverLining and senior advisor to the MCBP, which is led by atmospheric scientist Robert Wood and colleagues at the University of Washington. Wanser describes even more ways scientists are thinking of sustaining corals: they could be genetically modified, or otherwise bred to withstand warmer waters. Robust corals could also be moved into new areas and replanted. But doing these on ecosystem-wide scales would entail a tremendous undertaking just to restore tiny parts of the whole. "The Great Barrier Reef, that's like reinforcing the Rocky Mountains. It's massive."

"Essentially, it's heat stress that's killing corals," explains Wanser. "They're affected by other stressors. Heat compounds the other stressors. Heat makes acidity worse. It's a compound stressor. Heat is the mother of all stressors on corals, and it's just like there's a certain delta, and they start to go." She recalls a recent Ocean Studies Board meeting, where a scientist offered a grim prognosis: there's maybe twenty years left for 95 percent of the world's corals. Year by year, the maps went red. "If I didn't work on solar geoengineering, I would have had to leave, because it was very emotional because to see that. It's stunning."

What would brightening marine clouds actually look like? Essentially, it consists of engineering devices to spray seawater. "There's certainly some technical challenges to be overcome, but the basic process of just taking sea water and filtering it and then spraying it out, at submicron size, is not that difficult a technical challenge," Harrison says. His modeling results suggest that there would probably need to be some stations far offshore, beyond the edge of the continental shelf, which would require floating platforms or ships. This gets pricey. The maintenance costs would probably be the larger part, and the whole project could cost around $300 million. Expensive, but then again, the reef brings in an estimated $6 billion to the Australian economy. In Harrison's conception, you wouldn't want to brighten the clouds all the time, or even every summer. Rather, it would be done when the coral was at risk of bleaching, which would require about two weeks of forewarning in order to cool the water down to the maximum extent.

"But, I mean, there's some real unknowns here, right?" Harrison says. "Because no one's ever done any field work on this. So, it's quite unknown. You know, there's a general belief here that you can only target low-lying marine stratocumulus clouds, if they're already occurring. But then there's also quite a large body of evidence that shows that production on the reef influences the local climate, and influences cloudiness in an actual analog to what we're sort of thinking about." Essentially, coral can produce a chemical that makes clouds form—though this research on how reefs modify their own climate is in its early stages. His research team is interested, and worried, about how that will change as the reef bleaches, and whether there might be a positive feedback loop, wherein less clouds mean even more bleaching. "So, to some degree, we might be putting the system back towards where it was with aerosol production on the reef. But we really don't know, so I don't want to over emphasize that. And it's probably impossible for us to know, because we started monitoring the reef too late in human history."

Indeed, marine cloud brightening comes with a lot of unknowns—in part because cloud–aerosol interactions are not well understood in climate science, more generally. For a big-picture look at what climate models can and can't tell us, I spoke with Ben Kravitz, an atmospheric science professor at Indiana University who coordinates a project that compares geoengineering model simulations. He explains: "The climate system is inordinately complex. It's one of the more complicated systems that we know how to deal with. A great example of this is clouds. If you look out the window on an airplane, you can see clouds with all sorts of different structures. They're moving, some of them are a couple meters across, some of them are tens of kilometers across. Some of them are organized, some of them are not. Basically, you can't model all of that behavior in any single model, because we don't have the computational power. If there were a way to understand how clouds behave, in such a way that we could parameterize those behaviors and put them in models that we could actually run, that would solve some of the largest uncertainties in climate science."

Newer climate models are better at controlling for clouds' varying sensitivities to aerosols, so perhaps there will soon be better tools to get information about the effectiveness of marine cloud brightening.

But for that to happen, there has to be funding. Kelly Wanser of the Marine Cloud Brightening Project says that applied cloud-brightening research could actually help us understand some of these basic unknowns. However, potential funding organizations may see controlled outdoor field experiments into cloud–aerosol interactions as geoengineering related, because they have geoengineering applications. For the US-based project, their next step is to actually test the nozzle with seawater, which they would like to do on the California coast. The association with geoengineering has made it difficult to raise funding to actually build and test these nozzles. "I think we talked to all of the relevant government agencies who could support this, and essentially there's no one willing to say, 'We'll just do it as the cloud–aerosol basic science.' They're like, 'No, the cat's out of the bag, this is geoengineering. We would have to get approval.'" So, on one hand, there's a potential technique that could have global applications, as well as regional or local ones for particular marine ecosystems—but we don't know how well it would work, or what it would take to do it. On the other hand, it's been difficult to fund the research needed to get those answers because of the stigma of geoengineering.

Another aspect of marine cloud brightening that lies at the edges of scientific understanding is the teleconnections in the system—for example, how clouds in one place are connected to weather in another place. Anthony Jones, a climate modeler, has simulated regional solar geoengineering using stratospheric aerosols. His work has examined what happens when only certain parts of the system are modified. He tells me, "I think it scares me, the thought of doing marine cloud brightening." Because of all these weird teleconnections that we don't understand, I ask? "Yeah, the teleconnections. I've been looking at that a bit in some of our [stratospheric aerosol] simulations recently. So if you cool the North Pacific, you can actually shift the position of the jet stream … You get cold temperature on the western half of America, and warmer temperature on the eastern half of America," Jones explains. "The teleconnections are almost unavoidable, and if you can cool a certain area significantly, you are going to change the climate and the weather response." For this reason, Daniel Harrison thinks any attempt to use marine cloud brightening on a global scale would bring up major questions around governance: "If you want to cool the whole planet by doing marine cloud brightening, you know, some places are going to cool more than others. You're certainly going to alter, to some degree, global weather patterns. Maybe not that much, in the scheme of things, but it might not have to be very much to disadvantage some group of people living in some certain place, while advantaging everybody on the average." On the other hand, the concern about shifting weather patterns in remote places is less severe when it comes to brightening marine clouds over a reef, versus trying to modify temperatures globally with this technique. Those are two different goals. It would be better to consider something like area-specific marine cloud brightening for reefs to be a form of radical adaptation, rather than geoengineering.

So who cares about coral reefs, besides enamored children watching movies about clownfish? Coral reefs are not just a backdrop for colorful fish and exotic species. Reefs protect coasts from storms; without them, waves reaching some Pacific islands would be twice as tall. Over 500 million people depend on reef ecosystems for food and livelihoods.[1] Therefore, keeping these ecosystems functioning is a climate justice issue. Again, over 99 percent of corals would be wiped out at a two-degree-Celsius temperature rise, and perhaps 70 to 90 percent would be lost at 1.5 degrees.[2] And even if temperatures eventually stabilize at 1.5 degrees of warming a century or two from now, it's not known how well coral reef ecosystems would survive a temporary overshoot to higher temperatures.

Are we basically agreeing to give up on coral, and all the other animals and plants and unique forms of life in reefs, and the human economies and communities they support? On a societal level, it seems so. Many coral scientists, however, aren't willing to give up, though they oscillate between hope and despair, as ethnographer Irus Braverman finds in her research. In her book *Coral Whisperers*, Braverman describes how this polarity maps on to the rift between those conservation scientists who believe that it is possible to use traditional conservation methods (like withdrawal of human impacts), and others who take a more interventionist approach (on the basis that the natural systems are already fundamentally altered). Interestingly, she notes that "female scientists, many of them young and with diverse backgrounds, have taken the lead in promoting narratives of hope and models for assisted evolution."[3] But the restoration efforts in which many coral scientists are engaged stop short of intervention in the climate; some scientists in Braverman's book describe these efforts as holding together a patchy safety net, or as reef gardening—according tiny spaces of management that would hopefully survive an overshoot, like an outdoor aquarium in which to keep them until global warming is managed.

Is the survival of these life-forms and lifeways important enough to warrant research and discussion of geoengineering, or to justify coming up with a different conceptual category and language around geoengineering that would include more targeted interventions? It seems not. Nonhuman life is relatively absent from the anthropocentric geoengineering discussion, even though, as the saying goes, extinction is forever.

"The corals are a little bit like the canary in the gold mine," Harrison says. "They're very, very temperature sensitive. I really do think it's just a harbinger of things to come. You know, the coral ecosystem might collapse first, but I think there might be quite a few more ecosystems that'll follow it. I think that life is very resilient, but ecosystems as we know them aren't." Other ecosystems are also at high risk from even small changes in global mean temperature: Arctic ecosystems, mountain glaciers, and the Redwood forests in California, for instance. So are species that can't move quickly and find another suitable ecosystem. "It's the things that already live at the kind of extreme ends of the scale, and that can't move, right? So coral reefs, you know, they're stuck in already some of the warmest waters. If it gets too hot for them there, then (a) they can't move, and (b), they've got nowhere to go anyway. And the same with the extremely cold ecosystems. And the same with the Redwood forests. I guess. Trees can't up and move quickly enough to keep up with climate change."

Once you delve into the temporalities of the climate change problem—and especially the permanence of some of these changes, like extinction—it's easy to see how the idea of solar geoengineering makes its entrance. You're not reading this book in 1990, when carbon dioxide concentrations were still in the neighborhood of 350 parts per million. At this point, most people would agree that there's at least a chance we don't decarbonize before we lock in dangerous change—and for sensitive species and ecosystems like coral, the danger threshold has already been passed. The question, then, emerges: Can one use solar geoengineering to keep ecosystems on life support and forestall climate tipping points, while also decarbonizing?

Enter the "peak shaving" scenario, which uses solar geoengineering to "shave the peak" off of warming while carbon dioxide levels are being brought down. While we've discussed marine cloud brightening above, "solar geoengineering" in this context usually implies stratospheric aerosol injection, which would be a global-scale program.

In a nutshell, the most basic version of this peak-shaving scenario (depicted in Figure 2) means using specially designed high-altitude aircraft—perhaps a small airline's worth—to constantly fly aerosol precursors into the stratosphere. These aerosol precursors would cause the formation of particles made from sulfur, calcite, or some yet-to-be-determined substance. Why, if the world is trying to reduce particulate air pollution, would we put more particles up there? The particles are injected into the stratosphere—a layer of the atmosphere above where clouds form, and higher than planes usually fly. This means they would not fall back to earth in only a few days, as pollution from trucks and factories tends to. Rather, they would circulate around the whole planet, staying aloft for a year or so. Nevertheless, such an undertaking would not be without human health impacts. One study estimated an additional 26,000 deaths per year with enough sulfur-based geoengineering to offset one degree Celsius of warming, due to air quality and ultraviolet exposure; for comparison, 4 million people currently die each year from degraded air quality.[4] Indeed, the idea is to create a blanket

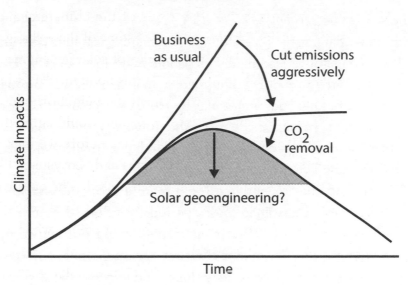

Figure 2. *Conceptual diagram of using solar geoengineering to "shave the peak" off a temperature overshoot. Sometimes called the "napkin diagram," based on a presentation by John Shepherd at Asilomar in 2010. Version source: Doug MacMartin*

of intentional, high-altitude pollution that would reflect something like 1 to 2 percent of incoming sunlight, perhaps less. Depending on one's perspective, this may sound mundane, or it may sound like an alarmingly unprecedented intervention into a poorly understood system.

We've come full circle, then, since the beginning of this book. Is "buying time" for carbon removal a legitimate reason for doing a limited amount of solar geoengineering? Or is it a weak justification for a project that will send the earth system careening down a dangerous road? Once started, how would people make sure the carbon really does get removed? In this chapter, we will look at some best-case and worst-case scenarios for a climate intervention program that includes solar geoengineering.

Kingston, Jamaica, July, 92°F / 33°C

Rain clouds hang in the Blue Mountains but never descend. The sidewalks in Kingston are jammed with people in business suits walking to work and street vendors selling sweets, making their way through fumes and honking and reggae blaring from car stereos. It is a glorious morning. I catch a ride up to the university, where there's a meeting about solar geoengineering research governance—the first such convening here in Jamaica.

The country has recently suffered from a drought, and while people here are accustomed to dry periods, yearlong droughts are something new. "Unfamiliarity is transforming our sensitivity into a vulnerability," explains the first speaker, climate scientist Michael Taylor. Jamaica has steep slopes and narrow coastal plains; with limited water storage, farms are largely dependent on rainfall. Livelihood, well-being, and water access here and elsewhere in the Caribbean are linked to the rains. But now the rain is more variable; the "nature" of rain is changing. Nighttime temperatures are soaring; one has to run the fan straight through the night. The climate will keep changing, Taylor says, with 98 percent of days here being "hot" by the 2090s—he emphasizes that only 2 percent of them will be cool.

It's sweltering in this room, and we're shifting uncomfortably on our wooden seats. There are a few fans, an adaptation to the broken AC, but even the local residents are sweating beads. Someone makes the inevitable joke about our prospects for controlling the global climate when we can't even control the climate in the room—a joke that's been made at approximately one-third of the geoengineering meetings I've ever been at. We press on.

Most people in the room are confronting the idea of geoengineering for the very first time, and their initial thoughts are diverse. A policymaker explains that he's read an article that suggested climate warming could create a whole new economic zone of mining and exploration—which would suit Northern countries, while those in the tropics would lose opportunities. Another person says that he's just installed solar panels to take his house off the grid: Will solar radiation management impact those? One speaker recalls a James

Bond movie he saw, in which someone controls an orbital laser beam. Someone else notes that manipulating the climate is like manipulating genes; it can be done in the wrong direction. An ethicist says that there's always some price to pay for engaging with technology, and we are in this problem because of technology. There's promise and peril; we are flying in the face of God.

But the conversation keeps returning to two themes. One is equity: it's the norm for Jamaica to be on the receiving end of things. Someone asks: "How can that be changed?" Jamaicans have regional alliances, they network, they negotiate as a block of Caribbean or small island developing states. But there's a disparity in terms of population when you're a small country. Discussion turns to this disparity in terms of historical emissions, and inequality also comes up in carbon trading, which can let polluters off the hook. "In an ideal world, we have iron-clad politics *before*" confronting something like this, one speaker asserts. Another asks: Can we afford *not* to look at an issue like this?

A second theme is capacity. A policymaker explains that the research system here was originally set up to train agricultural researchers for plantations, not to teach classical subjects. "Those of us in countries like Jamaica need to develop basic research," he says, because the problems people are trying to solve here may be different than those in other places. But here, they don't have the computing resources to run many computationally intensive climate models.

When it comes to designing a solar geoengineering program, both metaphorically ("program" as in a course of actions) and literally ("program" as in coding on a computer), a small developing country like Jamaica has a limited capacity to write it. The organization that pulled together the meeting, the Solar Radiation Management Governance Initiative, now coordinates a fund for researchers in developing countries, which provided an initial round of $430,000 toward eight projects that look at how solar geoengineering could impact things like droughts in Southern Africa or the spread of cholera in South Asia. This is an important step for the philanthropic sector, and its NGO and academic partners. Yet it's still a drop in the bucket compared to what would be needed for genuine inclusion in research design.

Algorithmic governance

Who *does* get to write the program for geoengineering? The verb "program" is rooted in the Latin *-graph*, connoting a written plan. Geoengineering would be a program to be authored, to be written, with real choices about what goes into the plan. In the end, though, program might not be the best metaphor, because it still has resonances of something "fixed" that you'd receive on a disk (even if these are now obsolete technology, as our software auto-updates). Solar geoengineering requires a more dynamic practice. "Responsive" or "adaptive" governance tries to connote this; but still, "responsiveness" seems like a tacked-on quality, something that modifies geoengineering after the fact, rather than being written into the fabric of what it is.

Let's follow these overlapping meanings of "programming," and also draw in another fuzzy term: "algorithm." "Algorithm," on a basic level, signifies a set of instructions; a recipe of sorts. Today, though, algorithms have taken on a broader meaning. They have become agents that determine aspects of our social reality: helping a billion-plus people get where they're going, assisting us in finding information, driving cars, manufacturing goods, assigning credit, shaping financial markets, and more, as science historian Massimo Mazzotti describes.[5] Wouldn't solar geoengineering inevitably be yet another one of these domains governed by algorithms, aided by an invisible computational hand?

We can be certain that aerosol geoengineering would be implemented in some kind of institutionalized program: with research milestones, perhaps "stage gates," flight operations, monitoring operations, and so on. Geoengineering with aerosols would also involve the use of computer tools to design a literal *program*—a recipe or operation—that would figure out the optimal way to put the particles in the stratosphere in order to achieve a combination of climate goals while minimizing negative impacts. In short, someone, somewhere, would write code for a system that would block a certain amount of sunlight, monitor the effects, and then adjust the system again. Researchers call this a "feedback control algorithm" because it would guide geoengineering using feedback from climate observations.

This is complex enough: now add the complexity of possibly using multiple climate engineering techniques. For example, a recent paper from a collaboration between scientists in China, India, and the United States simulated "cocktail geoengineering," which involved using two different geoengineering strategies— stratospheric aerosols and cirrus cloud thinning—to best restore preindustrial temperatures and precipitation.[6]

Then, add a temporal dimension: stratospheric aerosol geoengineering would likely take place over a time span of 150 years or more (at which point, if enough of our descendants make it through the twenty-first century, they will hopefully not only found better ways of removing carbon, but also improved upon our rusty old technology for deploying and monitoring the particles).

You can see how heavy computing would be crucial to a problem this complex. The resulting climate could be seen as some kind of human-machine-nature collaboration or dialogue, with constant back-and-forth retuning. In the parlance of the scientists working on it, it's about feedback and adjustment. Humans would input the goals. These could involve changing global temperatures, reducing sea level rise, stopping Arctic sea ice loss, or some other combination of ends that would likely be subject to long negotiations. It's quite possible that a set of decision rules for solar geoengineering could be created in a quasi-democratic matter—likely by the United Nations, where technical-expert delegations from various countries would hammer out the goals and the scheme for monitoring results, and so forth. But you can also very readily see the possibility that it will *not* be done that way—just consider how countries like Jamaica experience international decision-making processes.

Looking at solar geoengineering as an algorithm allows us to draw from the emerging literature on "algorithmic governance," which questions how algorithms are used to make decisions that pattern an increasing number of aspects of our lives. One key issue is about transparency and the black-boxing of algorithms: How can a geoengineering system be designed for openness and "algorithmic accountability"—that is, explainability in real time?

There is also the danger that bias could be coded into the program. This could happen because the underlying climate data is uneven; for example, war-torn countries are going to have gaps in their data. Or bias could be introduced due to variations in how problems are defined. Droughts, for example, can be hydrological, agricultural, or meteorological, each of which would be defined using different thresholds. Theoretically, the system could disadvantage vulnerable peoples without the explicit intention to do so, just because of poor data or poor problem definitions. (Of course, this is on top of the basic bias that determines who gets the education and power to even be in the position of writing computer code and making decisions.)

Given all the advances in computing, we might take for granted the ability to carry out the programming part. However, researchers point to varying constraints on this—both in terms of computing resources and qualified personnel. This isn't just the case for small island states or developing economies. Even the US public sector is stunningly constrained when it comes to running climate models; the computing infrastructure of the commercial cloud is soaring by comparison. Scientists I spoke to in several countries pointed to high-level expertise as another limitation. The labor time from qualified humans analyzing outputs is a constraint on geoengineering research (and on how advanced these algorithms can become), probably more so than computing resources. This is, of course, a matter of both training personnel and funding their research. Labor-time of qualified scientists seems like a worldwide constraint—though in theory, it should be solvable, given that politicians everywhere pay lip service to STEM training.

One challenge is to expand that training globally. In 2017, the geoengineering research program in China held the first geoengineering research course in Beijing for scientists from the developing world.

They provided model results for the entire earth system for students to analyze, because they believed students would best understand how to select the model parameters that were most important in their country. "It should not be some sort of teacher and student thing," clarifies project leader John Moore at Beijing Normal University, speaking about international collaboration more broadly. "It should be an equal relationship." He explains, "The international collaboration, and everybody being a sort of fair and equal partner, is a big priority from the Chinese viewpoint." China has no wish to do geoengineering unilaterally, and wants to avoid being seen as eager to geoengineer, he says. I ask him why he thinks that the Chinese are so interested in international cooperation. "They care about how the country's image is ... I guess it's sort of patriotism in a way, that people have a love of their country, and they don't want to be the international bad guy. They want to be the good guy, the nice guys. Given a choice, that's the natural choice."

International cooperation and collaboration, then, are part of a best-case scenario—one that many people are actively working toward. These cooperative-minded researchers and funders can help with capacity building. Still, they can't fully address the underlying structural inequalities between disparate working contexts. All this is to say that when thinking about the algorithm, we can't forget the material resources needed to make it work—both workers and infrastructure. Beginning an international, interdisciplinary research initiative *now* would create and hold the space for researchers to think more deeply about what transparent, democratic algorithm design might look like.

Keeping humans in the loop

Over the course of the next few decades, as we weigh climate intervention, machine learning and artificial intelligence will no doubt continue to make advances. What does it mean for these two capacities to evolve together? To be clear, scientists thinking about feedback control algorithms for solar geoengineering are *not* interested in mixing in artificial intelligence. Rather, they view these as systems where humans would be very much in the loop.

Ben Kravitz, the atmospheric science professor, points out two important problems with employing machine intelligence. "Number one, you have to actually believe what you're designing is correct." Understanding the underlying physical system, he tells me, is crucial. If you don't, and "you just say, 'Well, I don't really care, let's just wrap a controller around it and be done with it,' then everything could be fine until you get some weird—'weird' is I guess a technical term in this case—some weird behavior that you can't explain but that really messes things up. And Rumsfeld's 'unknown unknowns'. That's always a concern."

"Just because you can automate something doesn't mean it's a good idea to do so," Kravitz cautions. A second problem is that a machine intelligence might see an optimal outcome differently. "That is subjective—and not. ['Optimal'] is a really important word. If you're an economist and you can reduce everything down to dollar amounts, what you call 'optimal' might be different from what say a politician calls 'optimal,' because there are various additional concerns." Kravitz recalls the Three Laws of Robotics, by science fiction writer Isaac Asimov: First, a robot may not injure a human; second, a robot must obey the orders given by humans except where such orders would conflict with the First Law; third, a robot must protect its own existence, as long as that does not conflict with the first two laws. "That's sort of why they were invented, because what a machine calls 'optimal' is not necessarily what a human would call 'optimal.'" Kravitz points to the analogy of the Federal Reserve, a controller for a very complex system: "Depending on whether you call the Federal Reserve 'optimal' … it's basically a bunch of experts substituting for that machine, doing control theory on a poorly understood system." On the other hand, he notes, there can be problems with human decision processes, too. "Do you want the computer to do everything for you, or do you want people to be involved, even if that means reduced performance?"

To gain a better handle on this, I dropped in on control systems engineer Doug MacMartin at Cornell, who's authored numerous papers in top journals on potential designs of geoengineering systems (he has also collaborated with me on a project about how to incorporate community ideas into geoengineering research). MacMartin offers me some bread that he made, and begins to humor my questions about what it means when geoengineering and artificial intelligence grow up in the same time frame.

MacMartin, like Kravitz, emphasizes that deciding the goal of a climate intervention system is clearly a human activity: "That's a value, those are value judgments. Once you then say 'these are all the things that I care about,' you could essentially imagine that there's an algorithm that determines, given all the information that it has available to it and given the goals—here is the best thing to go do." Conceivably, some complicated deep-learning algorithm could help in this: one that has a much more advanced model of the climate system, and projects the future based on its knowledge of past climates and goals imposed by humans that tell it the performance metric. "'Here's what we care about. This matters this much, we don't want the rainfall here to deviate by more than this. We don't want this to change by more than that amount. And go find the solution in that space that is robust in all of your uncertainties about the current state of the system, and uncertainties about how the things evolve.' It sort of does the best job of balancing in this multidimensional goal space."

This is all hypothetical, of course, and I tell MacMartin that if this program existed, the outcome seems like it would be a human collaboration or negotiation with the program. I begin speculating: "It's so complex that you'd have to say, I care about Arctic ice, and I care about precipitation in this vulnerable region, and I care about XYZ—and then it would come up with something—but then that thing it comes up with would have this other thing that might cause a problem, and then you have to go—"

MacMartin asks: "You've done optimizations before, right?"

"Not really."

"This is the way they always work. Every optimization you ever do is like, 'Here's what I care about.' The computer then comes out and says, 'That's the optimum.' You look it and go, 'That's not what

I wanted.' You realize it's like, 'I only specified these variables, and I didn't specify this one over here, and it found a solution that never even occurred to me where it improves these, but destroys this thing over here.'"

I'm thinking that it sounds like a big mess—but this is what engineers deal with all the time, and a lot of our technological systems *do* actually work much of the time. "What you really want, in some sense," MacMartin explains, "is an iterative process ... And you presumably want a human in collaboration with that process who can then basically say, 'Wait a minute, that might have been what I asked for, but it's not what I wanted.'" MacMartin makes an important point—in the feedback class he teaches, they don't do anything that's optimal, because "optimal" is so tough to pin down. "You tend to do just as well by not optimizing it quite so much, if you know what I mean."

Do we get anywhere by looking at geoengineering as a program, or as software? "I think if we think about it as software, the first thing that comes to my mind is to think back to Star Wars [the Reagan-era missile defense program]. Which, as long as you thought about missile defense as a physics problem, seems solvable. The instant that you think about the missile defense system as a giant piece of software that happens to interface with physics, then you just laugh at it and say there's no way we could ever make this work."

MacMartin, like Kravitz, thinks that letting humans stray too far out of the loop would be risky: "I would say the biggest risk would be engineers being over confident in the ability of computer algorithms and allowing the computer too much leeway to make decisions. I don't personally buy into the idea that the computer eventually becomes sentient to prevent you from allowing you to turn it off. I think you can always turn it off." So, there will be no malicious artificial general intelligence, in MacMartin's view. But he identifies two other risks. The first is that something unexpected happens that is outside the training data that you've used to train a machine learning algorithm. The second is the risk that people become overreliant on the infrastructure and fail to understand the interconnected parts of the system.

In terms of social effects, like unemployment, the risks of these technologies developing together are probably indirect, MacMartin judges: "I think the bigger issue with them maturing at the same time is probably far more, on some level, related to the Trump factor—on steroids … Just as we look now and we think, 'Wow, George W. Bush. I wish we still had George W. Bush.' It wouldn't surprise me if in thirty years we say, 'I wish we had Trump.' Because if half the country is unemployed and unemployable forever, and there is no foreseeable pathway to get somebody who's forty years old to be employed in any meaningful way, that could have some pretty serious social repercussions—and combining that with something as powerfully upsetting for the human relationship with the universe as taking responsibility for the entire climate, and as inherently globalizing …" When people are struggling to find employment, he suggests, one reaction is to elect someone like Trump. "That nationalistic tendency is kind of at odds with the global implications of doing geoengineering." He pauses. "I suspect that issues have far less to do with narrowly how AI is being used in conjunction with geoengineering, but broadly in terms of how both AI and geoengineering affect the human relationship with the rest of the world in antagonistic directions. That could lead to really serious problems."

Designing the program

What is the best-case scenario for geoengineering? Scientists often laugh when I ask them that. "I would say the best-case scenario is if we figure out a way not to do it," replied Ben Kravitz, which is actually most people's first answer. "Barring that, I would say the best-case scenario is that we do it in an intelligent way, where we are designing geoengineering so that it will do what the politicians want it to do, minimizing side effects, and that there is an appropriate government structure … and just people working in a way that I believe they can work. Not the way they usually do." He adds, "We could go out and do geoengineering tomorrow, really poorly. That's what scares me."

Doug MacMartin has a similar view, sketching out a hypothetical scenario where the world manages mitigation that curbs warming to two and a half degrees, and uses solar geoengineering for a century to bring temperatures down to one and a half degrees. "We do that in a way where adjusting aerosol injection at different latitudes balances all sorts of different climate impacts. So that in principle, almost nobody on the planet is actually harmed [by geoengineering itself], and that there is strong international trust in whatever organizations are involved in making decisions, so that everybody on the planet feels that their voice has been heard, and accepts the fact that this limited deployment of solar geoengineering is better in some aggregate sense than not having had any geoengineering, and so it doesn't result in conflict."

"Geoengineers" are often caricatured as being in bed with fossil fuel companies, stranding us in the business-as-usual era; or otherwise, as gritty realists who lack the imagination to see the social transformation that's truly possible. Yet in the best-case articulation of many researchers, solar geoengineering *does* represent a rather-utopian dream. You have to believe that people are really capable of long-term thinking and cooperation to even articulate these scenarios—much less spend your time researching them. MacMartin adds that a best-case scenario, socially speaking, would frame geoengineering as a conscious acceptance of responsibility for the climate, rather than simply an effort to control it. "Accepting responsibility for it, I think, speaks to a maturation, and a way of expanding one's moral sphere further. Expanding one's moral sphere to the rest of the planet to future generations, to nonhumans, and saying we actually have responsibility for the betterment of all them. It's certainly possible that a hard discussion about solar geoengineering could push humanity in that direction."

While climate engineering researchers don't tend to interpret this question of a "best-case" climate scenario through the lens of a comprehensive, long-term solar geoengineering program, they do have a *sense* about what a climate engineering program might look like—contingent on what society is doing with mitigation and carbon removal. Ideally, solar geoengineering would be limited in scope. And it would be limited in time.

"Would a marginal deployment of geoengineering harm anybody?" Peter Irvine, an atmospheric scientist based at Harvard University, is trying to help answer this empirical research question. He explains that modelers often look at a scenario where all warming is offset, and evaluate who wins and loses in this extreme scenario. "Would offsetting 0.1 Celsius with geoengineering dial anything back, or would it amplify certain things? Does a marginal deployment help or hinder? ... How many more increments of cooling can you add before you start running into new issues that come with geoengineering?" Irvine and colleagues' work looks at what happens when just enough stratospheric aerosols are used to offset half of a doubling of CO_2 in the atmosphere for a hundred years. They find that using this smaller amount could avoid many of the previously reported impacts of aerosol geoengineering upon the hydrological cycle, such as extreme or decreased precipitation. With halved warming, everything seems to scale up without major hazards. "Beyond half ... some of the real differences start popping out a bit more." I asked Irvine about his best-case scenario for using solar geoengineering, and like most scientists, he says the best case is strong emissions cuts. "Against that backdrop ... I think some deployment carefully, carefully scaled up, little by little, over the course of a decade or two. ... Gradually halving the warming, possibly halting the rate of warming some decades in, against the backdrop of trying to cut emissions and then bringing them back down."

Even though Irvine is a climate scientist who thinks about the temporal aspects of a solar geoengineering system, he doesn't necessarily take a long-term, programmatic view. "You've got to think kind of decade by decade. Like, what do we do now, in this decade?" Irvine asks: "Who are we to say what the 2100 climate policy should be?" He thinks it is possible to have a climate policy that includes solar geoengineering but not negative emissions, where it would be decided decade by decade whether to ramp back the solar geoengineering or continue it, and points out that the same questions apply to making policy around negative emissions. "How quickly should we get to zero, and how quickly should we go negative? I think these are questions that, it's a bit daft to impose what we think… to basically, to meet some arbitrary 2°C target or 1.5°C target people in 100 countries agreed on in one meeting somewhere, and to assume that it's going to bind people 100 years from now."

Many scientists see solar geoengineering as a temporary measure that could be phased out if carbon removal was succeeding. Oliver Morton, in *The Planet Remade*, calls this temporary geoengineering scenario the "breathing-space approach," in that it allows incremental use of solar geoengineering to create breathing space for decarbonization.[7]

This intuition that solar geoengineering is a temporary intervention develops, I think, not only from an engineering perspective that thinks about resilient systems, but also from a moral sensibility. Kravitz says that solar geoengineering is not a permanent solution: "It's not something that we want to just do forever, or at least I don't, because I think too much can go wrong." Similarly, MacMartin thinks we'd eventually want to restore the climate using carbon removal, in order to have an exit strategy and avoid inflicting an implied commitment to solar geoengineering on future generations. Most stratospheric aerosol scenarios last 200 years, he says, and there's probably no deployment scenario that's less than a hundred years (with the caveat that it's possible that someone will find a really cheap way to pull CO_2 out of the atmosphere during that time period). Kelly McCusker, a climate modeler who has studied what happens when a solar geoengineering program is ended abruptly, also emphasizes that solar geoengineering will need to be done in combination with carbon dioxide removal or mitigation. "That's my feeling, but I don't necessarily have a good basis for that. I mean, in part, it's like you can look at these plots and just know it would be completely immoral to do it on its own."

The scientific consensus around this feeling is reflected in the IPCC's Special Report on 1.5°C, which explicitly assesses solar geoengineering in terms of its potential to limit warming to this amount "in temporary overshoot scenarios as a way to reduce elevated temperatures and associated impacts." It states that "if considered, SRM [solar radiation management] would only be deployed as a supplement measure to large-scale carbon dioxide removal."[8] This report actually reflects quite closely the views MacMartin, Kravitz, and others have already published, and their sense of how the practice might be best used, given that the aim of the report is to summarize the existing literature—and these researchers are the ones who wrote the existing literature. At the same time, this consensus on how it should be used might not be shared by politicians or industry leaders, who are often thinking on much-shorter timescales, and with different considerations in mind.

Why, exactly, is it so important that geoengineering be temporary? One issue is that of "termination shock," mentioned at the beginning of this book—the phenomenon in which if solar geoengineering was suddenly ceased, temperatures would shoot back up to a level commensurate with the greenhouse gas concentration of the atmosphere.

Some scientists argue that the risk of termination shock is not as high as portrayed in the media or academic literature. For a termination shock to happen, the geoengineering intervention would have to be large: as Oliver Morton writes, if solar geoengineering were "a relatively modest affair, the termination shock would be more a termination shudder."[9] Likewise, Andy Parker and Pete Irvine argue that a sudden termination is unlikely, and preventable.[10] Because solar geoengineering is cheap, they argue, it would take a pretty massive catastrophe to halt it and keep it turned off. For instance, 70 percent of GDP of the United States or China could be wiped out, and they could still deploy solar geoengineering for less than 1 percent of their post-catastrophe GDP. When it comes to such external forces, Irvine says: "If we're talking large-scale nuclear war, everything else that we're technologically dependent on is going to kill us first. I mean, the fact that I don't know how to grow food or catch food or hardly even look after myself … I think that's what's going to do us in, rather than a slightly greater warming off the back of your mushroom cloud–induced nuclear winter." It's true that people could *choose* to stop geoengineering, but Irvine finds this similarly unlikely, because for termination shock to be a big issue, you're already have to be decades into the program. "You're a generation into this. It's normal. It's as normal as, you know, irrigation and rivers. It's everyday … If you start to run through this story line, and put yourself in the perspective of a world that's thirty, forty years in … I think it would be as unthinkable as international shipping or aviation ending." Parker and Irvine also add that it would take months for a disruption to a solar geoengineering program to have any effect upon temperatures, because the aerosols would take months to thin out. "This is crucial for analysing the risks of termination shock, as it means that humanity would have a period of several months in which to resume deployment of SRM in the event of a disruption," they write. They point out that even if solar geoengineering was used to offset a large amount of warming, it could be slowly phased out over the course of decades without a shock. They also suggest criteria for a solar geoengineering system that would be robust against termination shock: it would need to be geographically distributed, affordable enough for multiple actors to maintain independent systems or backup hardware, and slow to lead to damages following disruption. "If back-up deployment hardware were maintained and if solar geoengineering were implemented by agreement among just a few powerful countries, then the system should be resilient against all but the most extreme catastrophes," they write.[11]

Other researchers, however, view the risk a bit differently. In one study, ecologist Christopher Trisos and colleagues modeled the consequences of solar geoengineering termination on other species.[12] Their study looked at the "climate velocity" of different species, which is the rate at which plants and animals have to move to keep their climate the same. Basically, if solar geoengineering was started and then stopped, many species would not be able to move quickly enough to keep up, threatening extinction for corals, mangroves, amphibians and land mammals. I caught up with Trisos one spring day in Washington, DC, to hear more about his research. "To the extent that risk is probability times consequence," he explained, "even if that probability is tiny, if the consequence is worse than anything we could conceive of over a similar time period of climate change without geoengineering; then for me, it's big enough of a risk to really think twice about the geoengineering discussion. Essentially, if we are exposing ourselves to that level of planetary risk, should we even think about geoengineering anymore in the first place?" Trisos thinks that the extreme severity of a termination shock, even if unlikely, raises the bar for geoengineering researchers to show that termination "is not a risk, or at least a low, low probability."

Atmospheric engineering on an ocean planet

There's another key reason that solar geoengineering would need to be nested within carbon removal: we live on an ocean planet. Ocean acidification has been called the "other CO_2 problem," and solar geoengineering wouldn't directly help ocean acidification much. (Solar geoengineering would likely affect ocean acidification through a host of other secondary effects—temperature effects on terrestrial biomass, hydrological cycle changes, changes in marine productivity[13] —but it doesn't directly address the issue of increased CO_2.)

Solar geoengineering may help mitigate sea level rise, though there are some limitations. Sea level rise has two drivers: ocean heating (because warmer waters expand), and melting ice. To stop the former, we'd have to address the energy imbalance in the ocean, which would mean returning it to roughly preindustrial conditions. Imagine that temperatures rose, but then we did geoengineering to lower temperatures back down half a degree, suggests Pete Irvine: this would reduce the direct atmospheric-driven, temperature-driven effects, such as heat waves. However, the ocean has layers. The surface ocean can respond to solar geoengineering in decades, but the deep ocean takes hundreds or thousands of years to respond. For example, if we reached two degrees Celsius and did geoengineering to come back to 1.5, the *rate* of heating would slow, but the ocean would not necessarily stop heating. "It's going to take thousands of years to reach a new balance."

By bringing temperatures down, one could stop ice from melting in places like Greenland, Irvine notes, which is otherwise going to experience a slow runaway feedback loop in which it loses mass over thousands of years. But in parts of Antarctica, on the other hand, staving off melting might not be possible, since Antarctica has points that become unstable with only a little bit of warming. "Because of the way the glaciers meet the ocean, when they start to retreat, they have kind of a runaway retreat. Again, very slow, like a couple of centuries. Five centuries. But once it starts, it's not a temperature-driven thing; it's a dynamic-driven thing … Once the ice shelf is sheared off or melted away, it's not there to hold the ice sheet back and there's this kind of dynamic response." Irvine explains that some ice may have crossed that tipping point already: "Quite big chunks of West Antarctica may have already crossed this threshold, and it might not be possible to dial them back. But there's other parts that may trigger a little later. Or may trigger much later …" It's not like everything is safe below 1.5 Celsius, and then there's a sudden step change when the world reaches 1.5. "I think there's lots of little mini steps. And the sooner you arrest warming, the fewer of these steps you cross." Some things are not reversible by solar geoengineering. Irvine continues: "You could imagine quite easily if that migrant species' last refuge gets pushed off the top of a mountain because it's getting too warm, well, it can't come back because it's dead. So I think there are examples where you could say, the atmosphere can respond back and we can dial it back. We can dial it back five decades later and get it back to where it was." But some aspects of the ice sheet response can't be dialed back. "And obviously if certain harms and damages have occurred, things have died, glaciers have melted away permanently … You know, there's certain things you can't take back."

Best-case solar geoengineering: A temporary measure for conservation?

One of the best reasons to consider solar geoengineering, in my view, could be to preserve species during an overshoot—but would this even be possible? I asked Christopher Trisos, the ecologist, about this. He's hesistant to say that geoengineering could save species, he says, because while some ecological impacts could be forecast, they really depend on the social choices that are made around geoengineering, and Trisos's impression is that "the scenario's uncertainty is still so big that ecologically, it is just anyone's game." There are areas that could be investigated, though, such as biome changes or wildfires, as well as disease vectors like mosquitoes, which we know a fair bit about. "What happens to the insects that are vectors for a lot of really nasty diseases like malaria, chikungunya, Zika? Is a geoengineered world better or worse for Zika in the Americas?" Or would it be worse for cholera in Asia? I wonder. It could be the opposite; maybe the conditions it sets in place climatologically are more preventative than promotive. We don't know. "A geoengineered world with potentially more Zika, more cholera, more malaria ... I'd much rather take climate change without geoengineering."

There's so much that's unknown, but to me, it's noteworthy that not very much research has been done on the ecologies of geoengineered worlds. I ask Trisos to speculate on this gap. He hypothesizes that ecologists don't want to lend the idea credibility. "The 30,000-foot view of what geoengineering could mean for ecosystems—my sense is a lot of people are reluctant to do that, or just not interested, because they view it more as the sci-fi fringe with the crazy people at climate conferences, and the real work of ecologists is to try and show how bad climate change can be. To promote greenhouse gas emission reductions. And also, I think in a normative way, geoengineering goes against what a lot of ecologists hold dear about promoting resilience of the planet, and natural recovery of ecosystems, and giving things space and time to have an adaptive capacity. The idea of putting your thumb on the thermostat of the planet is antithetical to that." The study of speculative futures under geoengineering is on the cutting edge of climate change ecology research, Trisos adds, which also makes it a tough sell for ecologists. "If you're going to push that research frontier forward, they would rather focus on a conventional climate change scenario than look at the more unconventional, potentially fringe ones like geoengineering."

Of course, actual conditions in a world with higher CO_2 and lower temperatures are unknown, and no one would have any reason to model it, unless they're thinking about geoengineering. For example, Trisos points to the possibility of a high-CO_2 world where woody vegetation encroaches on grasslands: Could the Serengeti be turned into a shrubby landscape that thickens into a forest under geoengineering? "In a lot of Southern Africa, where I'm from, savannas are really beautiful landscapes. They're important for eco-tourism. They're important for a lot of livelihoods around grazing and livestock. They have a lot of endemic species. They're ancient grasslands; fire has been a large part of the maintenance of those grasslands for hundreds of thousands of years. And if we turn down the temperature and have increased atmospheric CO_2, the extent to which that increases the invasion of these grassland and savanna areas by woody shrubs, I think, is important to know. It's potentially really concerning if you lose large swaths of grassland as a result of geoengineering. You could get bald patches. I mean, we haven't run the models yet. We don't know."

I understand the reluctance of ecologists to spend their time on the speculative imaginary of solar geoengineering when there's so much other essential research to do. Yet I worry about the idea of solar geoengineering moving into political discourse without these ecological aspects being investigated—for it is not only risks that could be ignored, but potential uses, too.

For Kelly Wanser of the Marine Cloud Brightening Project, the purpose of solar geoengineering is to keep ecosystems stable and intact, so that they don't start to break down in an unrecoverable way while the world gets its act together on climate change. I ask her about the best-case scenario for this. "A good scenario would be you use just the amount of solar geoengineering you need to keep all the coral reefs from disappearing," she answers. "To keep the ice sheets stable. To keep the methane trapped in the ground and things like that, for as long a period as you need in order to safely keep those systems stable, while you're bringing down the greenhouse gas concentration and you determine that you have a safe situation to proceed … The best-case scenario is, nothing really bad happens.

You use enough solar geoengineering to prevent the really devastating changes in the system, and you take as many measures as you can as quickly as you can to restore the underlying balance of forces in the atmosphere." She returns to the medical metaphor: "If you do it sooner and you keep heat levels from rising higher, you need less, and it would therefore be safer. Like any intervention. If we let temperatures continue to rise, then the counterforcing we're going to need is going to be pretty strong, and at that point you may also have other changes in the system going on."

Indeed, waiting a long time to do solar geoengineering, and then suddenly deciding to do it in the wake of an "emergency" like a massive drought, would not be the best-case program—because during that waiting period, the world is breezing past some irreversible tipping points, like species being senselessly lost forever. On the other hand, it is difficult to understate just how preliminary the idea of solar geoengineering is—basic crucial questions around how ecosystems may respond have scarcely been explored.

10

Reckoning

Downtown Los Angeles, January 2019, 27°C / 80°F

Empty playgrounds have a certain feeling. There is one in Pershing Square: a brand-new swing set with no children in sight, like a good intention fallen flat. Pershing Square is a public plaza in downtown Los Angeles whose abstract shapes, color scheme, and concrete modernism evoke both totalitarianism and the year 1992, like stepping into an MTV version of a public square in Turkmenistan. The square is mostly paved, overseen by a dark-purple tower.

It's high noon, and I feel like I've dropped into an alternate universe, or into some science fiction dystopia. There are bodies in the square, despite the metal bars on the benches cruelly designed to repel them. Everywhere, bodies are lying: in the shade of the dusty-orange metal spheres, or slumped around the perimeter of the fountain, perhaps sick, perhaps just sleeping in the noon heat.

To describe these people as bodies is reductive. The people here have lives and emotions and stories. For them, the plaza may not be as strange as I'm experiencing it to be: it may be a haven, an open space where they can stay peacefully during the daytime hours. Yet I don't want to normalize how these souls have come to inhabit this public space. Many of them are visibly in pain, and these streets are not a safe space for them. Los Angeles has about 59,000 homeless people, and 1,200 died on the streets in 2017 and 2018. To walk across the plaza at midday is to confront a society in which this is completely ordinary, in which the world moves around these bodies and the souls they contain, editing out any pain. It is like climate change: everyday and terrifying at the same time. My three-year-old is uncommonly quiet, and tugs on me, wanting to be carried through the square. I pick her up, thinking that whatever psychological mechanism allows most of us to walk through plazas like these is the same mechanism that allows us to walk past the signs of climate change.

Perhaps another reason I'm thinking about *bodies in the streets* in the park today is that there's a small group of people attempting disrupt this psychic mechanism, to de-normalize climate change, by placing their bodies in the street. About fifty people crowd around a black coffin in the far corner of the park, away from the other bodies. A woman with curly gray hair, who reminds me of my first-grade teacher, hands me a neon-green flier. It reads: "Sorry to bother you, but did you know that you are living during the sixth mass extinction?" They are like time travelers from a possible future, who have come back to warn us. Two people standing near the stoplight hold up a wide black banner with the words "Climate Emergency." The few passing pedestrians hardly look: the small gathering probably appears to be a religious organization, one of those quirky sects you see on street corners in big cities shouting about something arcane.

The protest has been organized by Extinction Rebellion, a direct-action movement that started in the UK. In cities all around the world on this day, Extinction Rebellion groups are staging die-ins. Here in Los Angeles, people are holding signs: "Where do you see yourself in twelve years?" The idea is that there are only twelve years left to stop climate change. The movement is "rebelling for life." My daughter climbs onto a bench and I give her some bunny-shaped crackers. She munches them morosely, gazing into traffic, apparently envigorated by protesting every aspect of daily life besides extinction. It's a diverse yet familiar group: black bandanas, young women with rosé-pink hair, silver-haired grandfathers, a guy in a black cowboy hat with a vaping device shaped like a bottle of Jack Daniels. "Biodiversity is being annihilated around the world," they say. "Our seas are rising. Flooding and desertification will leave vast tracts of land uninhabitable. ...The breakdown of our climate has begun. ...There will be more wildfires, unpredictable storms."

After the rally on the corner, they hoist up the coffin and carry it down the block, into a crowded food market. The procession marches past stalls of grass-fed beef, currywurst, wood-fired pizza, pupusas, a shop called "eggslut" where they sell fancy egg sandwiches, a ramen stall, and PBJ.LA, a restaurant dedicated to "elevated" peanut butter and jelly sandwiches. Through the crowds, they call: "We refuse to bequeath a dying planet to future genera-tions by failing to act now!" A few diners pause their eating to pull out their phones and take pictures of the coffin.

The thing is, some of the climate impacts they are chanting about may well have come from IPCC reports. None of the information they cite is out of step with the countless PhD-filled meetings I've been to on climate impacts. Their flier has *citations* on it, including figures to two decimal places. It mentions ice sheets, ocean acidifi-cation, and kelp forest death. "California started the water year with 44.3 percent of its area affected by drought." Et cetera. They're basi-cally right. It's not ignorable. And yet there is a capacity to ignore it, just like there's a capacity to ignore all the people in the plaza with bare feet and nowhere else to go.

I wrote this book because, like the people who showed up with their bodies to rebel for life, I'm gravely concerned about what the scientific evidence indicates for the future of life on earth. I'm worried that climate change will become so severe that even more people will suffer, and that in the midst of that suffering, people will grasp for solar geoengineering without adequate caution. I am worried that geoengineering will be used to protect material assets at the request of those who own those assets, without regard to vulner-able communities who lack any assets. This is what keeps me up at night. One seed of hope, though, is that many of these climate action groups didn't even exist when I started writing this book.

People often ask me, "What do you really think about solar geo-engineering?" I think it should be avoided, and at the very minimum be treated with great care. It's not yet a "thing"; it's still a notion, and we understand so little about it. There is a lot more research to be done. I think this research should be done, because the risks of climate change are so great. But I'm concerned that this research will be taken from the hands of the scientists who conduct it, and used or implemented by incompetent politicians or malicious regimes.

I also think that carbon removal technology should be pursued vigorously—and that without it, the push for solar geoengineering may be stronger. But I think carbon removal requires strong public advocacy and demand if it is to succeed at scale. This public support implies a tremendous shift in perception, and in values. It's not simply an engineering or infrastructure issue. Many of the new climate action groups, including Extinction Rebellion, Sunrise Movement, and the Climate Mobilization, among others, have expressed support for carbon drawdown. For instance, the Sunrise Movement currently calls for "funding massive investment in the drawdown of greenhouse gases" as part of its Green New Deal platform. Drawdown was part of Extinction Rebellion's demands, too, if sporadically. It is my hope that activist groups who grasp the climate situation can advocate for drawdown and help carbon removal become a reality, shaping it in ways that benefit rather than hurt communities around the world.

Taking account

We are just now beginning to reckon with climate change itself. The cognitive dissonance created by the coffin in the hipster food market, or the crowd in the park appearing like fanatics while reciting lines from scientific assessment reports, illustrates what it's like to be living at the moment of the shift. Here we are again, at the same point where we began this book: the shift, desperation point, where a critical mass of people realizes just how bad climate change will be, and how late we are to address it.

The desperation point could be used as a moment of exploitation, as in the logic of the "shock doctrine" that Naomi Klein traces in her work; it could take form as a crisis that consolidates power elsewhere. But it could also be a moment of reckoning, of taking account. "Accountability" first meant taking account, counting—it's linked with reckoning, etymologically. Writer and organizer Clare O'Connor writes about how at first, "accountability" had overtones of reckoning with God.[1] But during the eighteenth century, the term took on more positivist tones around sharing and evaluating; giving an account of something. Even today, though, "accountability" still conveys a moral sense.

Many critical thinkers perceive geoengineering as something that avoids accountability—a workaround for reckoning with not only the thermodynamics of climate change, but also their deeper implications and causes. As Kyle Whyte explains, colonial domination continues to be the problem that generates climate risks, to a large but not exclusive extent. Yet, Whyte observes, "geoengineering discourses isolate geoengineering as a topic and only add in colonialism, capitalist exploitation, imperialism and other forms of domination later as governance challenges or stakeholders' values or views that must be understood and weighed."[2] When it comes to geoengineering, scholars have thought about compensation for loss and damage; about transparency in research and decision making. But it's this narrative of giving an account, and moral sense of accountability for climate change itself, that are still missing, censored out, regarded as someone else's job.

Can there be a way of approaching geoengineering that considers the root causes of ecological degradation, and that weaves in accountability, reckoning, and reparation?

"Climate restoration" is an emergent language, though its limitations are immediately apparent. It is relationships that must be repaired and restored, not just the climate. As writer and social justice advocate Raj Patel and world historian Jason Moore explain:

> The rise of capitalism gave us the idea not only that society was relatively independent of the web of life but also that most women, indigenous peoples, slaves, and colonized peoples everywhere were not fully human and thus not full members of society. These were people who were not – or were only barely – human. They were part of Nature, treated as social outcasts – they were *cheapened*.[3]

Patel and Moore point to a host of *re-* words in their vision of "reparation ecology," whose "great promise is that humans – and what humans become – can thrive with the rest of the planet after the Capitolocene"; they discuss recognition, reparation, redistribution, reimagination, and recreation.[4] But not restoration: they caution that restoration of the environment after humans have damaged it is a flawed and backward-looking path. Rather, reparation ecology is a way to see both history and the future: "Redistributing care, land, and work so that everyone has a chance to contribute to the improvement of their lives and to that of the ecology around them can undo the violence of abstraction that capitalism makes us perform every day." With a similar focus on *re-* words, feminist science and technology theorist Donna Haraway suggests that our job "is to make the Anthropocene as short/thin as possible and to cultivate with each other in every way imaginable epochs to come that can replenish refuge."[5] She writes about replenishing refuges, about recuperation and recomposition—but describes these as partial, and stipulates that they "must include mourning irreversible losses."

In this talk of recognition, reparation, and replenishment, it's important to understand that a narrative account of reckoning or an apology for the present climate peril and all the processes of inequality and exploitation that led to it is not a closure, not a moving-on, but the beginning of long work ahead. Processes of reconciliation with indigenous groups around the world may be one place to look for inspiration, as well as for lessons about what is not helpful. Indigenous geographer Michelle Daigle has written about how Truth and Reconciliation Commission events in Canada become "confessional spaces of white guilt that were shaped by an ask-the-Indian dynamics where white people take up the majority of the space by asking Indigenous peoples what they could do to 'achieve' reconciliation and be a reconciled settler"—a dynamic that ends up putting the emotional labor on indigenous peoples.[6] She argues that this is "an era marked by the *spectacle* of reconciliation—a public, large-scale and visually striking performance of Indigenous suffering and trauma alongside white settler mourning and recognition—which secures, legitimates, and effectively reproduces white supremacy and settler futurity in Canada." Daigle points out that indigenous self-determination lies in the autonomy to remain unreconciled.

When it comes to climate justice, global reckoning, and repair of the climate as well as our relationships with each other and nonhumans, I don't know the contours of that road, the steps and missteps. I'm concerned that it will be easy for some to see this reckoning as threatening to the material success of carbon removal. In the US context, discussion of the violence that is often implicated in land use change or infrastructure, not to mention the question of whose lands these were and who took them, could appear to risk dismantling the coalitions that might be needed to get things started. I'm reminded of what Noah Deich of Carbon180 said about coalition building: there's the global Paris Agreement community, as well as energy, mining, and agriculture, all of whom need to embrace carbon removal, "not as a scary transformation for their business, but really the natural evolution for where they need to go to increase prosperity. To serve their customers, employees, shareholders, all of these key stakeholders better. It needs to come from the top down." Deich asks: How do business leaders hear the story? "How do they understand what the issues are, and who do they take the cues from? How do we get those people really putting carbon removal as the end zone, as the goal that we're in for collectively?" There won't be a single messenger, he intuits, but many people who will hear the story and message, and translate it into language that's resonant.

You can see a bipartisan vision here. In creating that discourse coalition of actors with disparate viewpoints and material interests, it's easy to see how things like reckoning, accountability, and acknowledgment of structural and physical violence could fall by the wayside (though I hope some of those top-down messengers will embrace these, too). People will say it's easier to generate momentum for carbon removal without going into this difficult terrain. But at the same time, just glossing over this fundamental need makes the end result less robust. It would be possible to continue with that cottage-level carbon removal, what we discussed as Level 1. But taking it net negative, to Level 2, will require such a dramatic social and political transformation that's hard to see it happening without this kind of deep social dialogue about injustice (both historical and contemporary). Embracing reckoning and accountability is not just a matter of doing what's morally right. It's also a pragmatic necessity for actually drawing down carbon.

Creativity in the ruins

When it comes to solar geoengineering, I think the best route forward is simply to go in and try to figure out as many of the unknowns as we can. For it is only in a few possible scenarios that I can see the specter of some of these ideas—stratospheric aerosol injection in particular—disappearing. One is a situation where mitigation is ramped up over the next decade and takes off in a wildly successful manner. A second is that a louder and more unified group of reputable climate scientists denounces solar geoengineering. This I find unlikely, because even though many mainstream climate scientists are terrified by the idea of geoengineering, they tend to support freedom to do research and find more things out. This is also a group that really grasps the risks of climate change, meaning that they may find further research warranted from the standpoint of climate risk management.

There's also a third possibility: that some other geoengineering, adaptation, or energy-related ideas come along that are seen as less risky, thus superceding the prospect of global stratospheric aerosol injections. For example, a high-engineering form of radical adaptation could be to engineer specific glaciers to keep them from melting. John Moore, glaciologist and leader of China's geoengineering research program, has recently been looking into this, and he wrote a comment with colleagues in *Nature* on the idea.[7] In Antarctica, there are two glaciers—Pine Island and Thwaites—that scientists have an especially nervous eye on. Warm ocean water flows in underneath them, something conventional wisdom says is unstoppable and irreversible due to the bedrock slope and geometry. But Moore suggests that building artificial islands to support the glacier ice shelf could act as an effective buttress.

Another technique would be to extract water from below the glaciers, in order to keep them from sliding off into the ocean. (Extraction of the water wouldn't work in Greenland, melting is on the surface—but Antarctica is different, with water coming in beneath the ice sheet and causing melting from below.) "You could actually remove a relatively small amount of water—like fifty cubic meters a second is what's produced underneath Pine Island," Moore tells me. One could make a tunnel in the bedrock that goes out into the ocean, thus draining this water and making the glacier sit harder on its bed. "You would slow the glacier actually by creating a few sticky spots. And, as the glacier slowed, it would reduce the [number] of icebergs going into the ocean." The advantage of this technique is that these two glaciers are relatively small. "In the case of Pine Island, it's forty kilometers across the mouth, and yet they will dominate Antarctica's contribution to sea level over the next two centuries," he says. "And, in a sense it's a very democratic, egalitarian way of dealing with sea level rise. Instead of trying to build walls around all the world's coastline, which actually means the rich countries will do it more than the poor countries of course, you can deal with the problem at the source where you have essentially something on the scale of a hundred kilometers to deal with instead of tens of thousands of kilometers of coastline to deal with."

I ask him if the engineering expertise exists to really do this.

"Actually, there are precedents. In terms of doing things like constructing a berm or something like that in front of the glaciers, you could look at things such as the construction of the Suez Canal or the building of Hong Kong's new airport. They needed on the order of a cubic kilometer of material to be moved. You can build a kind of artificial pinning point, like a little island. It doesn't need to go all the way to the surface, but you can build one of those for about a tenth of a cubic kilometer."

Of course, in Antarctica, he explains, you have to come to terms with the remoteness of the location and the damage to the pristine ecosystem from a big construction project: "This not something we envisage starting next year. This is something that can be done over a fifty-year time period. We could still reverse it. That's the thing.

And, when we've talked about this with glaciologists, there's a lot of horror at first." Antarctica, he notes, is a pristine international environment. "Clearly you're going to have to put some people there with a lot of stuff. That definitely will mess with the environment and the ecology, but if you compare the damage due to the collapse of the ice sheet, that's kind of dwarfed." The global sea level rise from these glaciers is staggering—on the order of meters per century. "You have to put that in perspective. A hundred million people moving from coastlines, compared with probably a few thousand people employed with building these devices." During the twentieth century, about a third of sea level rise came from Greenland, and very little came from Antarctica, because it's very cold there. But by 2100, Greenland is expected to contribute ten to twenty centimeters—and Antarctica could lend a meter.

This is a classic example of a hope-inducing engineering idea. But it still doesn't get around the need to cut emissions. As Moore points out, it targets only the sea level–rise piece of the problem: eventually, summer melting would occur on these glaciers, as it does in the Arctic. "All of these plans might come to nothing because of that. Essentially, we're trying to stop the ocean from doing the melting, but if the atmosphere's doing it as well, you're stuck. So, there has to be an exit strategy from the fossil fuel business-as-usual thing. This is not an alternative to mitigation by any means. It's just a kind of graceful way of trying to manage this unstable sea level rise. Depending on how we manage fossil fuels, that might end up with something where we could keep the ice sheets basically as they are now. Or if we don't stop the greenhouse gases, then we would probably just be in a case of trying to manage the rates of sea level rise. A sort of managed collapse, rather than a totally natural one."

Managing collapse implies softening deep losses. Indeed, reckoning with geoengineering in all its forms means coming to terms with loss; to explore what it means to "live in the ruins," in anthropologist Anna Tsing's phrase. Solar geoengineering comes as a shock to people who are attached to the idea of wilderness, who don't currently feel as if they are living in the ruins, and who haven't yet come to terms with the losses being experienced. In Beijing, though, where Moore lives, it's different—particularly with air pollution, the impacts of which people can see for themselves. "There's no denial," says Moore, "that somehow nature is out there, and it's pristine and beautiful. The fact is that everyone can see what we're doing. So, we've made this mess; we should clear it up. You can't rely on nature to do it."

For much of the world, the catastrophe is already here. Many are already living in the ruins, in the majority of the world, as well as in pockets of downtown Los Angeles. Kyle Whyte writes:

> What I think is ironic about a lot of geoengineering discourse is that the current situation that geoengineering is trying to preserve or save is actually a dystopia for some people. Members of dominant populations are trying to avoid their dystopia by preserving our dystopia. And that oftentimes is a very uninviting way to engage in a conversation because, for indigenous people, we often want to talk about the fact that we're still living in this situation where it's almost impossible for us to consent or dissent to anything. The legal policy barriers, the continued discrimination, and the massive habitat changes that have already occurred thanks to colonialism, capitalism, and industrialization—that's actually our starting point. Our starting point is not how do we maintain the current situation, but actually, how do we get out of it?[8]

Some of the techniques I've discussed in this book could be appropriated for the ends of reparation ecology, or for making the Anthropocene shorter—if we fully acknowledge the current situation, beyond just climate. Call it carbon removal, drawdown, regeneration, replenishment—it could shorten this dark time. If solar geoengineering is used—though I hope we can retire both the term and the idea—it, too, should be put in the service of this larger aim. Far from being a quick or easy "fix," climate restoration is an idea that is not just technologically optimistic, but also radically socially optimistic, for it implies that we could chart our way through this mess and reconstitute a climate that is safe for people to grow regenerative food and live healthy lives. The kind of lives many progressives imagine—where we are gardening, living in reasonably sized and energy-efficient homes, riding our bicycles, feeling healthy, and healing our relationships with each other and with other forms of life—are more likely to be possible at the end of the century, for our descendants, if we pursue multiple methods of carbon removal.

We need more thought experiments into radical adaptation, as well as new ideas for carbon removal—whether that means engineering glaciers, putting biochar into roadways, robotically farming kelp, turning carbon into rock beneath the earth, or other approaches yet to be discovered. We need creativity, both technologically and socially; to think beyond the boxes of capitalist economics, on one hand, and binary formulations, on the other. What other bold ideas might be out there? What forms of social organization will help them blossom? People will accuse new ideas involving technology of planting false hopes. But the hope doesn't inhere in the technology —it inheres in the people who would craft it: the workers, designers, collaborators, educators, engineers, cultural producers, farmers, and others. In the long term, they are the place to source hope—not in technology. Climate is a long game. The work of shaping these practices and technologies, and what comes after geoengineering, will also be long.

Acknowledgments

F irst and foremost, my gratitude to everyone who spoke with me for this project for your time, creativity, and energy.

I wouldn't have even thought to write this book without the suggestion from Andreas Malm. Many thanks to Rosie Warren at Verso, and thanks also to CM for the thoughtful edits.

At UCLA, Peter Kareiva is a phenomenal champion of books and writers: thank you. Daniel Swain's communication of climate science is an inspiration.

This book is the result of a decade of contemplating geoengineering, and it owes a debt to many people around the globe. My research collaborator Ilona Mettiäinen deserves special thanks for her curiosity, pragmatism, and generosity. I'm very grateful to Doug MacMartin as well as Charles Geisler, Sunny Power, and Steve Hilgartner at Cornell: thank you for your spirit of intellectual openness and inquiry.

Thanks to the community of geoengineering researchers. Special thanks to Ted Parson for all the dynamic conversations, to Ben Kravitz and Pete Irvine for your continued willingness to explain scientific concepts, and to Michael Thompson for all the links. I'm grateful for the critical global discussions with Shinichiro Asayama, Katharina Beyerl, George Collins, Olaf Corry, Jane Flegal, Oliver Geden, Clare Heyward, Drew Jones, Jane Long, Sean Low, Nils Markusson, Nils Matzner, Duncan McLaren, Matt Kearnes, Juan Moreno-Cruz, David Morrow, Oliver Morton, Simon Nicholson,

John Noel, Andy Parker, Christopher Preston, Jesse Reynolds, Kate Ricke, Dan Sanchez, Stefan Schäfer, Karolina Sobecka, Pablo Suarez, and Gernot Wagner: and my apologies to those whom I missed.

I'm also grateful for my fellow travelers in the writing life: Wendy Saul, Elizabeth Guthrie, Ellie Andrews, J. P. Sapinksi, Elisabeth Olson, Florian Mosleh, and Mercury. Finally, deepest gratitude to Laura Watson for the timeless care work.

Notes

Introduction

1. Vaclav Smil, *Energies: An Illustrated Guide to the Biosphere and Civilization*, Cambridge, MA: MIT Press, 1999, 5.
2. Ibid, 11.
3. The figure cited for current radiative forcing since 1750 is from the *5th IPCC Assessment Report* 2015, so it's probably a bit higher by the time you read this.
4. B. H. Samset et al., "Climate Impacts from a Removal of Anthropogenic Aerosol Emissions," *Geographical Research Letters* 45, 2018.
5. IPCC, "Summary for Policymakers," in *Global warming of 1.5°C: An IPCC Special Report,* eds. V. Masson-Delmotte et al., Geneva: World Meteorological Organization, 2018, 16.
6. Felix Creutzig et al., "Beyond Technology: Demand-Side Solutions for Climate Change Mitigation," *Annual Review of Environmental Resources* 41, 2016, 173–98.
7. IPCC, "Summary for Policymakers."
8. John Sterman and Linda Booth Sweeney, "Understanding Public Complacency about Climate Change: Adults' Mental Models of Climate Change Violate Conservation of Matter, *Climatic Change* 80:3–4, 2007, 213–38.
9. Pak-Hang Wong, "Maintenance Required: The Ethics of Geoengineering and Post-Implementation Scenarios," *Ethics, Policy and Environment* 17:2, 2014, 186–91.
10. Jeremy Scahill, "Leading Marxist Scholar David Harvey on Trump, Wall Street, and Debt Peonage," *Intercept*, theintercept.com, Jan. 21, 2018.

11. David Graeber, *The Utopia of Rules: On Technology, Stupidity, and the Secret Joys of Bureaucracy,* New York: Melville House, 2015, 120.

12. Ibid., 146.

13. McKenzie Wark, "What if This Is Not Capitalism Any More, but Something Worse?," NPS Plenary Lecture, APSA 2015, Philadelphia, PA, in *New Political Science* 391, 2017, 58–66.

14. Claire Colebrook, "What is the Anthropo-Political?" In *Twilight of the Anthropocene Idols*, eds. Tom Cohen, Claire Colebrook, J. Hillis Miller, London: Open Humanities Press, 2016, 86.

15. Andreas Malm, "For a Fallible and Lovable Marx: Some Thoughts on the Latest Book by Foster and Burkett," *Critical Historical Studies* 42, 2017, 267–75.

16. Matthew T. Huber, "Hidden Abodes: Industrializing Political Ecology," *Annals of the American Association of Geographers* 1071, 2017, 151–66.

17. Jesse Goldstein, *Planetary Improvement: Cleantech Entrepreneurship and the Contradictions of Green Capitalism*, Cambridge, MA: MIT Press, 2018, 14.

18. Leigh Stanley Phillips, *Austerity Ecology & The Collapse-Porn Addicts: A Defence of Growth, Progress, Industry, and Stuff*, Alresford, UK: Zero Books, 2015.

19. Nick Srnicek and Alex Williams, *Inventing the Future: Postcapitalism and a World without Work*, London and New York: Verso, 2016, 146.

20. Laboria Cuboniks, "Xenofeminism: A Politics for Alienation," laboriacuboniks.net.

21. ETC, *The New Biomasters: Synthetic Biology and the Next Assault on Biodiversity and Livelihoods*, ETC Group Communiqué 104, 2010.

22. Naomi Klein, *This Changes Everything: Capitalism vs. the Climate*, New York: Simon & Schuster, 2014.

23. Joshua Horton and David Keith, "Solar Geoengineering and Obligations to the Global Poor," in *Climate Justice and Geoengineering: Ethics and Policy in the Atmospheric Anthropocene*, ed. Christopher J. Preston, London: Rowman & Littlefield, 2012.

24. Jane Flegal and Aarti Gupta, "Evoking Equity as a Rationale for Solar Geoengineering Research? Scrutinizing Emerging Expert Visions of Equity," *International Environmental Agreements* 181, 2018, 45–61.

25. Kyle Powys Whyte, "Indigeneity in Geoengineering Discourses: Some Considerations," *Ethics, Policy and Environment*, 21:3, 2019.

26. Charles Eisenstein, "We Need Regenerative Farming, not Geoengineering, *Guardian*, Mar. 9, 2015, theguardian.com.

27. ETC, *The New Biomasters*.
28. Biofuelwatch, *Smoke and Mirrors: Bioenergy with Carbon Capture and Storage BECCS*, biofuelwatch.org.uk, 2015.
29. Jack Stilgoe, *Experiment Earth: Responsible Innovation in Geoengineering*, New York: Routledge, 2015, 8.
30. Anne Pasek, "Provisioning Climate: An Infrastructural Approach to Geoengineering, in *Has It Come to This? The Promise and Peril of Geoengineering on the Brink*, eds. J. Sapinski, H. J. Buck, and A. Malm, Princeton, NJ: Rutgers University Press, forthcoming.
31. Bent Flyvbjerg, Nils Bruzelius, and Werner Rothengatter, *Megaprojects and Risk: An Anatomy of Ambition*, Cambridge, UK: Cambridge University Press, 2003.
32. Ben Marsh and Janet Jones, "Building the Next Seven Wonders: The Landscape Rhetoric of Large Engineering Projects," in S. D. Brunn, ed., *Engineering Earth*, New York: Springer, 2011.
33. Flyvbjerg et al., *Megaprojects*.
34. David Nye, *Consuming Power: A Social History of American Energies*, Cambridge, MA: MIT Press, 1997.
35. Brad Allenby, "Infrastructure in the Anthropocene: Example of Information and Communication Technology," *Journal of Infrastructure Systems* 10, 2004, 79–86.

1 Cultivating Energy

1. Christy Borth, *Pioneers of Plenty: The Story of Chemurgy*, Indianapolis: The Bobbs-Merrill Company, 1942.
2. William Hale, *Farmward March: Chemurgy Takes Command*, New York: Coward McCann, 1939, 141.
3. Randall Beeman, "'Chemivisions': The Forgotten Promises of the Chemurgy Movement." *Agricultural History* 68:4, 1994, 26.
4. Mark Finlay, "The Failure of Chemurgy in the Depression-Era South: The Case of Jesse F. Jackson and the Central of Georgia Railroad," *The Georgia Historical Quarterly* 81:1, 1997, 78-102.
5. Beeman, "Chemivsions," 32.
6. Quentin R. Skrabec, *The Green Vision of Henry Ford and George Washington Carver: Two Collaborators in the Cause of Clean Industry*, Jefferson, NC: McFarland & Co, 2013, 193.
7. David Constable, "Green Chemistry and Sustainability." *In Quality Living Through Chemurgy and Green Chemistry*, ed. Peter K. Lau. Berlin: Springer, 2016, 2.

8. Tito Kuol, "Looking Downstream: The Future of Nile River Politics," *Harvard Political Review*, April 3, 2018, harvardpolitics.com.

9. Jon Abbink, "Dam Controversies: Contested Governance and Developmental Discourse on the Ethiopian Omo River Dam," *Social Anthropology* 20:2, 2012, 125–44.

10. See, for example, Oakland Institute, *Understanding Land Investment Deals in Africa, Country Report: Ethiopia*, Oakland, CA, 2011.

11. Jon Abbink "'Land to the Foreigners': Economic, Legal, and Socio-cultural Aspects of New Land Acquisition Schemes in Ethiopia, *Journal of Contemporary African Studies* 29:4, 2011, 526.

12. Lorenzo Cotula, "The International Political Economy of the Global Land Rush: A Critical Appraisal of Trends, Scale, Geography and Drivers," *Journal of Peasant Studies* 39:3–4, 2012, 666.

13. Philip McMichael, "The land grab and corporate food regime restructuring," *Journal of Peasant Studies* 39.3-4, 2012, 690.

14. Klaus Deininger, "Challenges Posed by the New Wave of Farmland Investment," *Journal of Peasant Studies* 38: 2, 2011, 218.

15. Nizar Manek, "Karuturi Demands Compensation from Ethiopia for Failed Land Deal," *Bloomberg*, September 21, 2017.

16. See Benjamin Niemark, S. Mahanty, and W. Dressler, "Mapping Value in a 'Green' Commodity Frontier: Revisiting Commodity Chain Analysis," *Development and Change*, 472, 2016, 240–65.

17. Jim Lane, "In for a Penny, In for a Pound: The Advanced Bioeconomy and All the Pivots," *Biofuels Digest*, December 28, 2017.

18. Carol Hunsberger, Laura German, and Ariane Goetz, "'Unbundling' the Biofuel Promise: Querying the Ability of Liquid Biofuels to Deliver on Socio-economic Policy Expectations," *Energy Policy* 108, 2017, 791–805.

19. Ibid.

20. Tania Murray Li, "After the Land Grab: Infrastructural Violence and the 'Mafia System' in Indonesia's Oil Palm Plantation Zones," *Geoforum* 96, 2018.

21. L. J. Smith and M. S. Torn, "Ecological Limits to Terrestrial Biological Carbon Dioxide Removal," *Climatic Change* 118, 2013, 89–103.

22. Patrick Moriarty and Damon Honnery, "Review: Assessing the Climate Mitigation Potential of Biomass," *AIMS Energy* 5:1 2017.

23. Timothy Searchinger, Tim Beringer, and Asa Strong, "Does the World Have Low-Carbon Bioenergy Potential from the Dedicated Use of Land?," *Energy Policy* 110, 2015, 434–46. See also Joseph Fargione et al., "Land Clearing and the Biofuel Carbon Debt," *Science* 319:6857, 2008.

24. Warren Cornwall, "Is Wood a Green Source of Energy? Scientists Are Divided," *Science*, January 5, 2017.

25. Eric Roston, *The Carbon Age: How Life's Core Element Has Become Civilization's Greatest Threat*, New York: Walker & Co., 2008.

26. David Biello, "Whatever Happened to Advanced Biofuels?," *Scientific American*, May 26, 2016, scientificamerican.com.

27. Stephen Mayfield, "The Green Revolution 2.0: The Potential of Algae for the Production of Biofuels and Bioproducts," *Genome* 56, 2013, 551–5.

28. Ann C. Wilkie et al., "Indigenous Algae for Local Bioresource Production: Phycoprospecting," *Energy for Sustainable Development* 15, 2011, 365–71.

29. Colin M. Beal et al., "Integrating Algae with Bioenergy Carbon Capture and Storage (ABECCS) Increases Sustainability," *Earth's Future* 6, 2018.

30. I. Ajjawi, et al., "Lipid Production in *Nannochloropsis gaditana* Is Doubled by Decreasing Expression of a Single Transcriptional Regulator," *Nature Biotechnology* 35, 2017, 647–52.

31. Kelsey Piper, "Silicon Valley Wants to Fight Climate Change with These "Moonshot" Ideas," *Vox*, Oct. 26, 2018, vox.com.

32. Natalie Hicks et al., "Using Prokaryotes for Carbon Capture Storage," *Trends in Biotechnology* 351, 2017, 22–32.

33. Mathilde Fajardy and Niall Mac Dowell, "Can BECCS Deliver Sustainable and Resource Efficient Negative Emissions?," *Energy and Environmental Science*, 10, 2017, 1389–1426.

2 Cultivating the Seas

1. Seasteading Institute, "Project Oasis," seasteading.org/project-oasis.

2. D. Krause-Jensen and C. M. Duarte, "Substantial Role of Macroalgae in Marine Carbon Sequestration, *Nature Geoscience* 9, 2016, 737–42.

3. Adam D. Hughes et al., "Biogas from Macroalge: Is It Time to Revisit the Idea?," *Biotechnology for Biofuels* 5, 2012, 86.

4. Rasmus Bjerregaard et al., *Seaweed Aquaculture for Food Security, Income Generation and Environmental Health in Tropical Developing Countries*, Washington, DC: World Bank Group, 2016, documents.worldbank.org.

5. Ibid.

6. Robert D. Kinley et al., "The Red Macroalgae *Asparagopsis taxiformis* Is a Potent Natural Antimethanogenic That Reduces Methane

Production during In Vitro Fermentation with Rumen Fluid," *Animal Production Science* 56:3, 2016, 282–9.

7. Howard Wilcox, "The Ocean as Supplier of Food and Energy," *Experientia* 38, 1982, 32.

8. Antoine de Ramon N'Yeurt et al., "Negative Carbon via Ocean Afforestation," *Process Safety and Environmental Protection* 90, 2012, 467–74.

9. Stuart W. Bunting, *Principles of Sustainable Aquaculture: Promoting Social, Economic and Environmental Resilience*, New York: Routledge, 2013.

10. E. J. Cottier-Cook et al. "Policy Brief: Safeguarding the Future of the Global Seaweed Aquaculture Industry," Hamilton, ON: UNU-INWEH and SAMS, 2016.

11. Sander W. K. van den Burg et al., "The Economic Feasibility of Seaweed Production in the North Sea," *Aquaculture Economics and Management* 203, 2016, 235–52.

12. Hughes et al., "Biogas from Microalgae."

13. Xi Xiao et al., "Nutrient Removal from Chinese Coastal Waters by Large-Scale Seaweed Aquaculture," *Nature Scientific Reports* 7:46, 2017, 616.

14. Cottier-Cook et al., "Policy Brief."

15. Calvyn Ik Kyo Chung, F. A. Sondak, and John Beardall, "The Future of Seaweed Aquaculture in a Rapidly Changing World," *European Journal of Phycology* 524, 2017, 495–505.

16. Krause-Jensen and Duarte, "Macroalgae in Marine Carbon Sequestration."

17. Jason Zhang et al., "Big Picture Resilience via Ocean Forests," presentation, ASCE Innovation Conference, 2016,

18. Alastair Bland, "As Oceans Warm, the World's Kelp Forests Begin to Disappear," *Yale Environment 360*, November 20, 2017.

3 Regenerating

1. Matthew Kearnes and Lauren Rickards, "Earthly Graves for Environmental Futures: Techno-burial Practices," *Futures* 92, 2017, 48–58.

2. The Land Institute, "Kernza Grain: Toward a Perennial Agriculture," landinstitute.org/our-work/perennial-crops/kernza.

3. L. Hunter Lovins et al., *A Finer Future: Creating an Economy in Service to Life*, Gabriola Island, BC: New Society Publishers, 2018, 61.

4. Brian Barth, "Carbon Farming: Hope for a Hot Planet," *Modern Farmer*, March 25, 2016.

5. Keith Paustian et al., "Climate-Smart Soils," *Nature* 532, 2016, 49–57.

6. David D. Briske et al., "The Savory Method Can Not Green Deserts or Reverse Climate Change," *Rangelands* 355, 2013, 72–4.

7. Rebecca Lave, "The Future of Environmental Expertise," *Annals of the American Association of Geographers* 1052, 2015, 244–52.

8. Lovins et al., *A Finer Future*, 178.

9. Charles Eisenstein, *Climate: A New Story,* Berkeley, CA: North Atlantic Books, 2018, 180.

10. Ibid., 181.

11. National Research Council, *Climate Intervention: Carbon Dioxide Removal and Reliable Sequestration*, Washington, DC: National Academies of Science, 2015.

12. Anne-Marie Codor et al., *Hope Below Our Feet: Soil as a Climate Solution,* Global Development and Environment Institute Climate Policy Brief 4, 2017.

13. Paustian et al., "Climate-Smart Soils."

14. Salk Institute, "Harnessing Plants Initiative," salk.edu/harnessing-plants-initiative.

15. Paustian et al., "Climate-Smart Soils."

16. See T. Thamo, and D. J. Pannell, "Challenges in Developing Effective Policy for Soil Carbon Sequestration: Perspectives on Additionality, Leakage, and Permanence," *Climate Policy* 16:8, 2016 and Pete Smith, "Soil Carbon Sequestration and Biochar as Negative Emission Technologies," *Global Change Biology* 22, 2015, 1315–24.

17. Smith, "Soil Carbon Sequestration."

18. Kristen Ohlson, *The Soil Will Save Us: How Scientists, Farmers, and Foodies are Healing the Soil to Save the Planet*, New York: Rodale, 2014.

19. National Academies of Sciences, Engineering, and Medicine. *Negative Emissions Technologies and Reliable Sequestration: A Research Agenda*. Washington, DC: The National Academies Press, 2019, 80.

20. Julia Rosen, "Vast Bioenergy Plantations Could Stave Off Climate Change and Radically Reshape the Planet," *Science*, Feb. 15, 2018, sciencemag.org.

21. See R. A. Houghton, "The Emissions of Carbon from Deforestation and Degradation in the Tropics: Past Trends and Future Potential," *Carbon Management* 45, 2013, 539–46; and Timothy M. Lenton, "The Potential for Land-Based Biological CO2 Removal to Lower Future Atmospheric CO2 Concentration," *Carbon Management* 11, 2010, 145–60.

22. Simon Evans, "World Can Limit Warming to 1.5C 'without BECCS,'" *Carbon Brief*, 2018, carbonbrief.org.

23. Detlef Van Vuuren et al., "Alternative Pathways to the 1.5°C Target Can Reduce the Need for Negative Emission Technologies," *Nature Climate Change* 8, 2018, 391–7.

24. Kim Naudts et al., "Europe's Forest Management Did Not Mitigate Climate Warming," *Science* 351, 2016, 6273.

25. Brendan Mackey et al., "Policy Options for the World's Primary Forests in Multilateral Environmental Agreements," *Conservation Letters* 82, 2015, 139–47.

26. National Research Council, *Climate Intervention*, 2015.

27. Gabiel Popkin, "How Much Can Forests Fight Climate Change?," *Nature*, Jan. 15, 2019, nature.com.

28. W. Sunderlin et al., "How Are REDD+ Proponents Addressing Tenure Problems? Evidence from Brazil, Cameroon, Tanzania, Indonesia, and Vietnam," *World Development* 55, 2013, 37–52.

29. Jon Unruh, "Tree-Based Carbon Storage in Developing Countries: Neglect of the Social Sciences," *Society and Natural Resources: An International Journal*, 24:2, 2011, 185–92.

30. K. Suiseeya and S. Caplow, "In Pursuit of Procedural Justice: Lessons from an Analysis of 56 Forest Carbon Project Designs," *Global Environmental Change* 23, 2013, 968–79.

31. Kristen Lyons and Peter Westoby, "Carbon Colonialism and the New Land Grab: Plantation Forestry in Uganda and Its Livelihood Impacts," *Journal of Rural Studies* 36, 2014, 13–21.

32. Eric Lambin and Patrick Meyfroidt, "Global Land Use Change, Economic Globalization, and the Looming Land Scarcity," *PNAS*, 2011, pnas.org.

33. E. Mcleod et al., "A Blueprint for Blue Carbon: Toward an Improved Understanding of the Role of Vegetated Coastal Habitats in Sequestering CO_2," *Frontiers in Ecology and the Environment* 7, 2011, 362–70.

34. C. Nellemann et al., *Blue Carbon: A Rapid Response Assessment* 78, United Nations Environment Programme, GRID-Arendal, 2009.

35. Ibid.

36. Sophia Johannessen and Robie Macdonald, "Geoengineering with Seagrasses: Is Credit Due where Credit Is Given?," *Environmental Research Letters* 11, 2016.

37. Shu Kee Lam et al., "The Potential for Carbon Sequestration in Australian Agricultural Soils Is Technically and Economically Limited," *Nature Scientific Reports* 3, 2013, 2179.

38. Kearnes and Rickards, "Earthly Graves."
39. Bronson W. Griscom et al., "Natural Climate Solutions," *PNAS* 114:44, 2017, 11645–50.

4 Capturing

1. Elizabeth Kolbert, "Can Carbon Dioxide Removal Save the World?," *New Yorker*, November 20, 2017.
2. Joelle Seal, "SaskPower's Carbon Capture Future Hangs in the Balance," CBC News, November. 23, 2017, cbc.ca.
3. IEA, *20 Years of Carbon Capture and Storage: Accelerating Future Deployment*, Paris:OECD/IEA, 2016; Juho Lipponen et al., "The Politics of Large-Scale CCS Deployment," *Energy Procedia* 114, 2017, 7581–95.
4. Glen Peters and Oliver Geden, "Catalysing a Political Shift from Low to Negative Carbon," *Nature Climate Change* 7, 2017, 619–21.
5. IEA, *20 Years of Carbon Capture and Storage.*
6. Greenpeace, *Carbon Capture Scam CCS: How a False Climate Solution Bolsters Big Oil*, Washington, DC: Greenpeace USA, 2015.
7. Cesare Marchetti, "On Geoengineering and the CO2 Problem," *Climatic Change* 1, 1977, 59–68.
8. Bryan Maher, "Why Policymakers Should View Carbon Capture and Storage as a Stepping-stone to Carbon Dioxide Removal," *Global Policy* 91, 2018, 102–6.
9. Alfonso Martinez Arranz, "Carbon Capture and Storage: Frames and Blind Spots," *Energy Policy* 82, 2015, 249–59.
10. IEA, *20 Years of Carbon Capture and Storage.*
11. Maher, "Carbon Capture and Storage as a Stepping-stone."
12. IPCC Working Group III, *Special Report on Carbon Dioxide Capture and Storage*, Cambridge and New York: Cambridge University Press, 2005, 442.
13. Alfonso Martinez Arranz, "Hype among Low-Carbon Technologies: Carbon Capture and Storage in Comparison," *Global Environmental Change* 41, 2016, 124–41.
14. NASEM, *Negative Emissions Technologies.*
15. Mikael Roman, "Carbon Capture and Storage in Developing Countries: A Comparison of Brazil, South Africa and India," *Global Environmental Change* 21, 2011, 391–401.
16. Arranz, "Hype among Low-Carbon Technologies."
17. Karin Bäckstrand et al., "The Politics and Policy of Carbon Capture

and Storage: Framing an Emergent Technology," *Global Environmental Change* 21, 2011, 275–81.

18. Philip J. Vergragt et al., "Carbon Capture and Storage, Bio-energy with Carbon Capture and Storage, and the Escape from the Fossil-Fuel Lock-in," *Global Environmental Change* 21, 2011, 282–92.

19. See, for example, James Gaede and James Meadowcroft, "Carbon Capture and Storage Demonstration and Low-Carbon Energy Transitions: Explaining Limited Progress," in T. Van de Graaf et al., eds., *The Palgrave Handbook of the International Political Economy of Energy*, London: Palgrave MacMillan, 2016.

20. APS, "Direct Air Capture of CO2 with Chemicals: A Technology Assessment for the APS Panel on Public Affairs," Washington, DC: American Physical Society, 2011.

21. Jeff Tollefson, "Sucking Carbon Out of the Air Is Cheaper than Scientists Thought," *Nature* 558, 2018, 173.

22. The Global CO2 Initative, "Global Roadmap for Implementing CO2 Utilization," 2016, globalco2initiative.org/opportunity.

23. Jocelyn Timperley, "Q&A: Why Cement Emissions Matter for Climate Change," *Carbon Brief*, September 13, 2018, carbonbrief.org.

24. Niall Mac Dowell et al., "The Role of CO2 Capture and Utilization in Mitigating Climate Change," *Nature Climate Change* 74, 2017, 243–9.

25. Ibid.

26. Stuart Haszeldine, "Can CCS and NET enable the continued use of fossil carbon fuels after CoP21?" *Oxford Review of Economic Policy* 32:2, 2016, 304–322.

27. Kathryn Yusoff, "Epochal Aesthetics: Affectual Infrastructures of the Anthropocene," *e-flux*, 2017, e-flux.com.

5 Weathering

1. Juerg M. Matter et al., "Rapid Carbon Mineralization for Permanent Disposal of Anthropogenic Carbon Dioxide Emissions," *Science* 352:6291, 2016, 1312–14.

2. Andy Skuce, "'We'd have to finish one new facility every working day for the next 70 years': Why Carbon Capture Is No Panacea," *Bulletin of Atomic Sciences*, October 4, 2016.

3. Ilsa B. Kantola et al., "Potential of Global Croplands and Bioenergy Crops for Climate Change Mitigation through Deployment for Enhanced Weathering," *Biology Letters* 134, 2017.

4. See "Theme 3—Applied Weathering Science," lc3m.org/research/theme-3/.

5. David J. Beerling, "Enhanced Rock Weathering: Biological Climate Change Mitigation with Co-Benefits for Food Security?," *Biology Letters* 134, 2017.

6. Paul Hawken, ed. *Drawdown: The Most Comprehensive Plan Ever Proposed to Reverse Global Warming*, New York: Penguin Books, 2017.

7. David P. Edwards et al., "Climate Change Mitigation: Potential Benefits and Pitfalls of Enhanced Rock Weathering in Tropical Agriculture," *Biology Letters* 134, 2017.

8. Ibid.

9. Kantola et al., "Potential of Global Croplands."

10. Edwards et al., "Climate Change Mitigation."

11. Ibid.

12. Henry Fountain, "How Oman's Rocks Could Help Save the Planet," *New York Times*, April 26, 2018, nytimes.com; Evelyn M. Mervine et al., "Potential for Offsetting Diamond Mine Carbon Emissions through Mineral Carbonation of Processed Kimberlite: An Assessment of De Beers Mine Sites in South Africa and Canada," *Mineralogy and Petrology* 112, 2018, 755–65.

13. See "Remineralize the Earth," remineralize.org.

14. L. L. Taylor et al., "Simulating Carbon Capture by Enhanced Weathering with Croplands: An Overview of Key Processes Highlighting Areas of Future Model Development," *Biology Letters* 134, 2017.

15. Ibid.

16. F. J. Meysman and F. Montserrat, "Negative CO2 Emissions via Enhanced Silicate Weathering in Coastal Environments," *Biology Letters* 134, 2017.

6 Working

1. Kathi Weeks, *The Problem with Work: Feminism, Marxism, Antiwork Politics, and Postwork Imaginaries*, Durham, NC: Duke University Press, 2011, 182.

2. Imre, Szemanand and Maria Whiteman, "Future Politics: An Interview with Kim Stanley Robinson," *Science Fiction Studies* 312, 2004.

3. Giorgios Kallis, *In Defense of Degrowth: Opinions and Manifestos*. Uneven Earth Press, 2018.

4. Ibid., 21.

5. Ibid., 12.

6. Michael Spectre, "The First Geo-vigilante," *New Yorker*, October 18, 2012, newyorker.com.
7. Whyte, "Indigeneity in Geoengineering Discourses," 10.
8. Lave, "The Future of Environmental Expertise."
9. "Firm to Perform Ocean Experiment," *BBC News*, June 21, 2007, news.bbc.co.uk.
10. See Fred Turner, *From Counterculture to Cyberculture*, Chicago: University of Chicago Press, 2006.
11. See "Regenerative Organic Certified," regenorganic.org.
12. Neera Singh, "The Affective Labor of Growing Forests and the Becoming of Environmental Subjects: Rethinking Environmentality in Odisha, India," *Geoforum 47*, 2013, 189–98.
13. Karen Pinkus, "Carbon Management: A Gift of Time?," *Oxford Literary Review* 321, 2010, 51–70.
14. Adam Greenfield, *Radical Technologies: The Design of Everyday Life*, London and New York: Verso Books, 2017.
15. Ibid., 147.
16. Nori, *Nori White Paper*. Version 3.0.1, February 19, 2019. Nori.com.
17. Imre Szeman, "Entrepreneurship as the New Common Sense," *South Atlantic Quarterly* 114:3, 2015.
18. Katja Grace et al., "When Will AI Exceed Human Performance? Evidence from AI Experts," *Journal of Artificial Intelligence Research* 62, 2018.
19. Srnicek and Williams, *Inventing the Future*.
20. Luke Dormehl, *Thinking Machines: The Quest for Artificial Intelligence and Where It's Taking Us Next*, New York: Penguin Random House, 2017.
21. Thomas Davenport and Julia Kirby, *Only Humans Need Apply: Winners and Losers in the Age of Smart Machines*, New York: HarperCollins, 2016.
22. Judy Wajcman, "Automation: Is It Really Different This Time?," *British Journal of Sociology* 681, 2017.
23. Stefan Helmreich, "Blue-Green Capital, Biotechnological Circulation and an Oceanic Imaginary: A Critique of Biopolitical Economy," *BioSocieties* 2, 2007, 287–302.
24. Elizabeth Johnson, "At the Limits of Species Being: Sensing the Anthropocene," *South Atlantic Quarterly* 1162, 2017.
25. Stewart Brand, *Whole Earth Discipline: Why Dense Cities, Nuclear Power, Transgenic Crops, Restored Wildlands, and Geoengineering Are Necessary*, New York: Penguin Books, 2009.
26. Ibid., "We Are as Gods," *Whole Earth Catalog*, Winter 1998, wholeearth.com.

7 Learning

1. Timothy Mitchell, *Carbon Democracy: Political Power in the Age of Oil*, London and New York: Verso, 2011.
2. Oliver Morton's discussion of interference in the nitrogen cycle is an excellent deep-dive into this. Oliver Morton, *The Planet Remade: How Geoengineering Could Change the World*, Princeton and Oxford: Princeton University Press, 2015.

8 Co-opting

1. Gavin Bridge and Philippe Le Billon, *Oil*, Malden, MA: Polity, 2013, 136.
2. Ibid.
3. Mike Berners-Lee and Duncan Clark, *The Burning Question*, London: Profile Books, 2013, 87.
4. Paul Griffin, *CDP Carbon Majors Report 2017*, 2017.
5. Kevin Sack and John Schwartz, "Left to Louisiana's Tides, a Village Fights for Time," *New York Times*, February 24, 2018, nytimes.com.
6. Bridge and Le Billon, *Oil*, 66–7.
7. Charles McConnell, keynote address, CO2 & ROZ Conference, Midland, Texas, December 3, 2018.
8. Dmitry Zhdannikov, "'Under Siege,' Oil Industry Mulls Raising Returns and PR Game," *Reuters*, January 24, 2019, reuters.com.
9. Stephanie Anderson. *One Size Fits None: A Farm Girl's Search for the Promise of Regenerative Agriculture*, Lincolin: Nebraska University Press, 2019.
10. Joel Wainwright and Geoff Mann, *Climate Leviathan*, London and New York: Verso Books, 2018, 30.
11. Marco Armeiro and Massimo de Angelis, "Anthropocene: Victims, Narrators, and Revolutionaries," *South Atlantic Quarterly* 1162, 2017.
12. Nina Power, "Demand," in *Keywords for Radicals*, Oakland: AK Press, 2016.
13. Berners-Lee and Clark, *The Burning Question*, 95.

9 Programming

1. OECD, *Marine Protected Areas: Economics, Management and Effective Policy Mixes*, Paris: OECD Publishing, 2017.

2. IPCC, *Special Report on Global Warming of 1.5°C.*

3. Irus Braverman, *Coral Whisperers: Scientists on the Brink*, Oakland: University of California Press, 2018.

4. Sebastian D. Eastham et al., "Quantifying the Impact of Sulfate Geo-engineering on Mortality from Air Quality and UV-B Exposure," *Atmospheric Environment* 187, 2018, 424–34.

5. Massimo Mazzotti, "Algorithmic Life," *Los Angeles Review of Books*, January 22, 2017, lareviewofbooks.org.

6. Long Cao et al., "Simultaneous Stabilization of Global Temperature and Precipitation through Cocktail Geoengineering," G*eophysical Research Letters* 44:14, 2017.

7. Morton, *The Planet Remade.*

8. IPCC, *Special Report on Global Warming of 1.5°C*, chapter 4, 55.

9. Morton, *The Planet Remade.*

10. Andy Parker and Peter J. Irvine, "The Risk of Termination Shock from Solar Geoengineering," *Earth's Future* 8:2, 2018, 249.

11. Ibid.

12. C. H. Trisos et al., "Potentially Dangerous Consequences for Biodiversity of Solar Geoengineering Implementation and Termination," *Nature, Ecology and Evolution* 23, 2018, 475–82.

13. Phillip Williamson and Carol Turley, "Ocean Acidification in a Geo-engineering Context," *Philosophical Transactions of the Royal Society A* 3701974, 2012.

10 Reckoning

1. Clare O'Conner, "Accountability," in *Keywords for Radicals*, Oakland: AK Press, 2016.

2. Kyle Powys Whyte, "Indigeneity in Geoengineering Discourses: Some Considerations," *Ethics, Policy and Environment* 21:3, 2018.

3. Raj Patel and Jason W. Moore, *A History of the World in Seven Cheap Things: A Guide to Capitalism, Nature, and the Future of the Planet*, Oakland: University of California Press, 2017, 24.

4. Ibid., 207.

5. Donna Haraway, *Staying with the Trouble: Making Kin in the Chthulu-cene*, Durham, NC: Duke University Press, 2016.

6. Michelle Daigle, "The Spectacle of Reconciliation: On the Unsettling Responsibilities to Indigenous Peoples in the Academy," *Environment and Planning D: Society and Space* OnlineFirst, 2019.

7. John Moore et al., "Geoengineer Polar Glaciers to Slow Sea-Level

Rise," *Nature* 555, 2018, 303–5.

8. Kyle Powys Whyte, "Geoengineering and Indigenous Climate Justice: A Conversation with Kyle Powys Whyte," in *Has It Come to This? The Promise and Peril of Geoengineering on the Brink*, eds. J. Sapinski, H. J. Buck, and A. Malm, Princeton, NJ: Rutgers University Press, forthcoming.

Index